ANDERSONIAN LIBRARY
WITHDRAWN
FROM
LIBRARY
STOCK
UNIVERSITY OF STRATHCLYDE

D1345727

INTERNAL GRAVITY WAVES IN THE OCEAN

Marine Science

Series Editor

Donald W. Hood
Institute of Marine Science
University of Alaska
Fairbanks, Alaska

Volume 1 **Bioactive Compounds from the Sea**
Edited by Harold J. Humm
and Charles E. Lane

Volume 2 **Internal Gravity Waves in the Ocean**
Jo Roberts

Additional Volumes in Preparation

INTERNAL GRAVITY WAVES IN THE OCEAN

Jo Roberts
Institute of Marine Science
University of Alaska
Fairbanks, Alaska

MARCEL DEKKER, INC., *New York*

MARCEL DEKKER, INC.

270 Madison Avenue, New York, New York 10016

LIBRARY OF CONGRESS CATALOG CARD NUMBER: 74-78969

ISBN: 0-8247-6226-6

Current printing (last digit):
10 9 8 7 6 5 4 3 2 1

PRINTED IN THE UNITED STATES OF AMERICA

To Tom, Lynneth, Evan, and Eiluned

PREFACE

This book was written as a summary of work, theoretical and observational, which has been done on internal gravity waves in the ocean. It is not a textbook, although some topics are prefaced by a discussion of elementary ideas and definitions. In writing the book, I had three audiences in mind: First, investigators of a particular field of internal wave study who may find some useful references; second, students who are presented (as I was) with the problem of studying internal waves and need some idea of where to turn for help; and third, scientists who, for some reason or other, would like to know something about internal gravity waves in the ocean. Some knowledge of classical mechanics and of differential equations is assumed but, except in Chap. 3, the mathematical developments are minimal.

Throughout, I have given more attention to recent papers, because they usually incorporate the results of older ones. This is not always true, of course, and the possibility must remain that some new and basic idea is buried in an obscure paper. In some cases, the same work of an author has been published in several forms, e.g., in a thesis and as a journal article. In referring to such work, I have attempted to use the reference most easily available. The novice should be aware, however, that sometimes the original version of the work, a thesis, a technical report, a talk at a symposium, may have illuminating comments and derivations which are omitted from the more available journal article. A reference for the original form of a work, if it has been published, is often given in the bibliography of the journal article. I have excluded a few papers from discussion because they apparently contain errors. No doubt they also contain good results; nonetheless, I did not attempt to isolate the valid from the invalid conclusions. To those authors whose names are not mentioned in this book, I apologize. Their work was generally of a specialized nature and had been already included in other review articles. Anyone interested in a particular internal wave topic, lee waves, for example, would do well to consult the review articles which I have cited on that topic.

Terms which are defined in the body of the text are underlined. A list of symbols follows the appendix. I have used the words "internal waves" and "internal gravity waves" interchangeably. A three-dimensional, rectangular coordinate system is assumed throughout. Vectors are denoted by superposed arrows and as the sum of the various components times the appropriate unit vector $\hat{i}$, $\hat{j}$, or $\hat{k}$. Thus, the radius vector may be written as $\vec{r}$ or $\hat{i}x + \hat{j}y + \hat{k}z$. Subscripts x, y, and z denote the respective components of a vector, e.g., $\vec{k} = \hat{i}k_x + \hat{j}k_y + \hat{k}k_z$. The magnitude of a vector is denoted by the vector symbol without an arrow: $|\vec{k}| = k$. Differentiation is always written out (that is, it is never denoted by primes or subscripts).

Two books are so basic to the study of internal waves that they deserve to be mentioned at the outset: Krauss' (1966b) *Interne Wellen* and Phillips' (1966) *Dynamics of the Upper Ocean*. I have drawn heavily from both. Krauss' book is an excellent place to begin one's study of internal waves. The developments are refreshingly lucid, the German is straightforward, and English summaries for each section are provided at the back of the book. Phillips' work, on the other hand, is more advanced and is, as Dr. Thorpe has commented, "full of interesting ideas." Indeed, it is a book impressive for both its insight and its foresight.

This book could not have been written without the cooperation and kindness of many scientists who patiently explained their work and other aspects of internal waves to me: Professors F. P. Bretherton, C. S. Cox, R. E. Davis, N. P. Fofonoff, W. Krauss, L. Magaard, J. W. Miles, W. H. Munk, O. M. Phillips, C. I. Wunsch;

Drs. T. H. Bell, Jr., B. Johns, E. C. LaFond, Y.-H. Pao, H. Perkins, and many more. Still others sent reprints, references, and manuscripts. Throughout the book I have done little more than reorganize the work of others, hoping all the while that the original intentions of the authors remain intact.

I am particularly indebted to Professor B. Haurwitz, whose lectures and subsequent conversations clarified many points about internal waves that had been bothering me for some time; to Dr. A. D. Kirwan, who suggested this project and supplied encouraging remarks at appropriate moments; and to my husband, Professor T. D. Roberts for his critical reading of the manuscript and the correction of many errors in physics.

Carol Campbell drew and plotted the map of internal wave observations, James Burton prepared most of the other figures, Annette Kokrine typed the manuscript and Mauricette Nicpon typed the final, camera-ready copy for publication; I am grateful to them all.

I am indebted to Cambridge University Press for their permission to use photographs from the *Journal of Fluid Mechanics* for Figs. 1-2, 1-8, 1-17, and 1-18; and to the American Meteorological Society for permission to use photographs from the *Journal of Atmospheric Science* for Figs. 1-19 and 1-20.

Support for this research was provided by the Office of Naval Research under Contract No. N00014-67-A-0317-0002-MOD-4, and by the Institute of Marine Science at the University of Alaska.

Fairbanks
May, 1973

Jo Roberts

CONTENTS

CHAPTER 1

OBSERVATIONS AND ANALYSIS OF INTERNAL WAVES

1.1 MAP AND TABLES OF OBSERVATIONS

A map showing the locations of reported internal wave observations is given in Fig. 1-1; the references for the map are given in Table 1-1. The areas into which the map is divided are quite arbitrary; their only use is to help locate a particular reference. No strict criteria were used in selecting the references. By and large, if the author reported observing internal waves, or even fluctuations of temperature or other oceanic parameters which might be construed as being caused by internal waves, the observation is included on the map.

Table 1-2 gives a summary of the observations by map area, with lakes listed separately. One should be cautioned about using the information in this table: Any one entry is not a compendium of all the observations which have been made for that area. Thus, for example, there have been four observations of diurnal internal tides reported for Area V, but only two of these give an estimate for the vertical displacement (Piip, 1969, gives 10 m; and Seiwell, 1939, gives 100 m). Since some areas include several different basin configurations with widely differing density distributions, this summary is probably of limited value. It would be more interesting to see a comparison of observations from different basins of similar form and/or density profiles. Such a comparison has

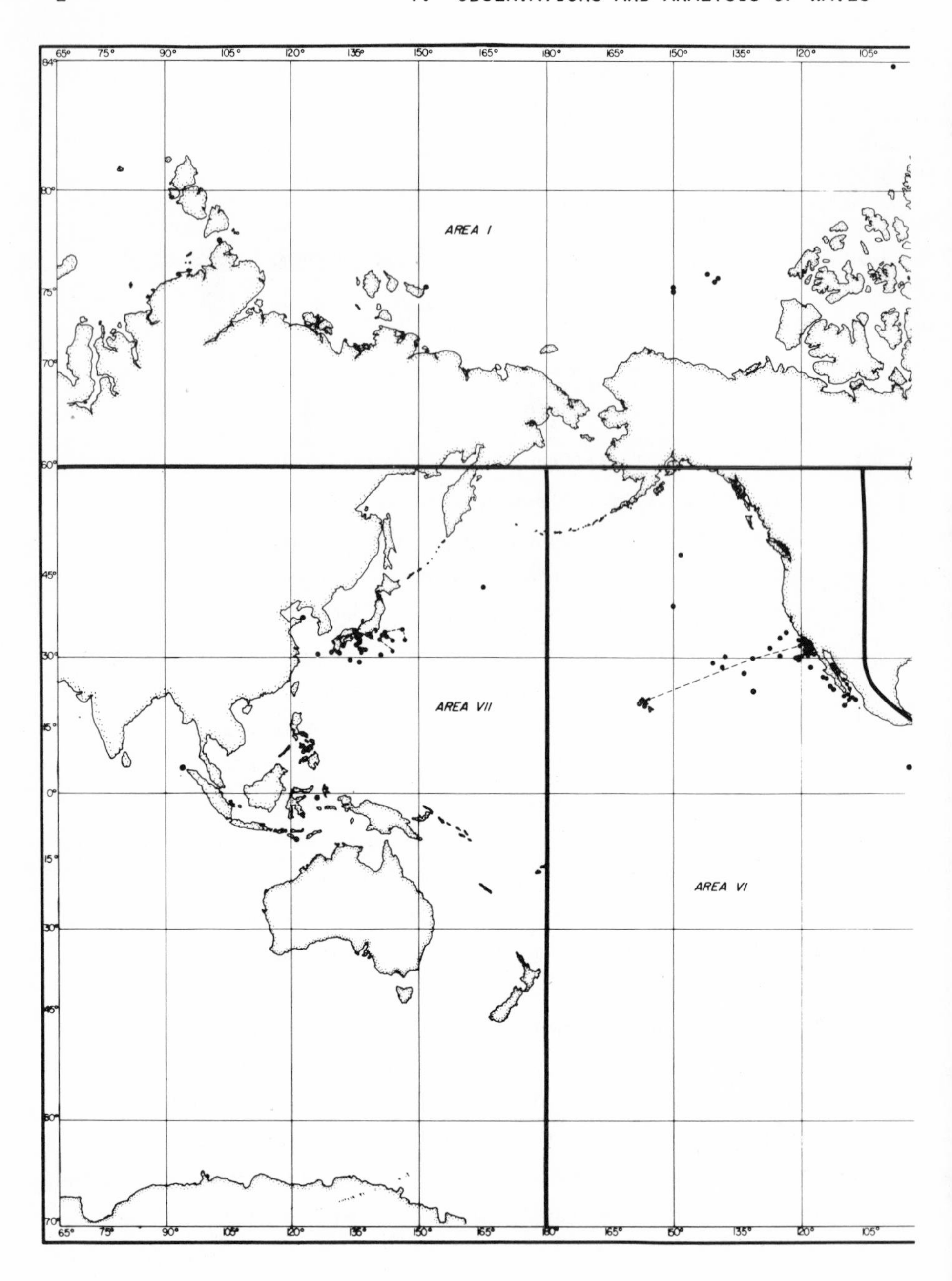
AREA I
AREA VII
AREA VI
65° 75° 90° 105° 120° 135° 150° 165° 180° 165° 150° 135° 120° 105°
84° 80° 75° 70° 60° 45° 30° 15° 0° 15° 30° 45° 60° 70°

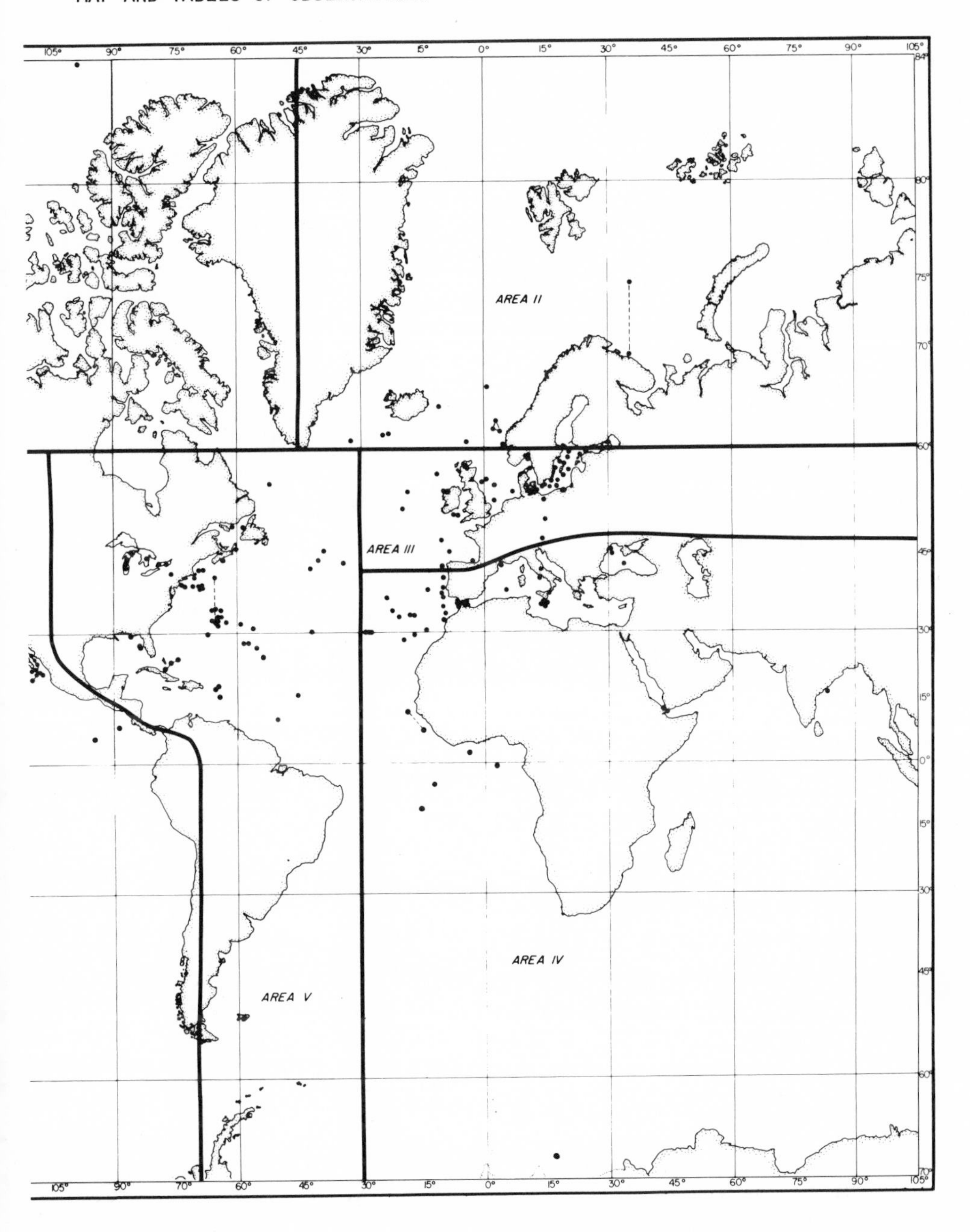

FIG. 1-1. Location of reported observations of internal waves.

TABLE 1-1

Key to Map of Internal Wave Observations, Fig. 1-1

Reference	Approximate Location
Area I	
Bernstein and Hunkins (1971)	Arctic Ocean
Fjeldstad (1936)	76°N, 153°E
Morse and Smith (1972)	Arctic Ocean
Neshyba *et al.* (1972)	85°N, 97°W
Yearsley (1966)	Arctic Ocean
Area II	
Fjeldstad (1952, 1964)	Norwegian Coast
Helland-Hansen (1930)	Faeroe-Shetland Channel
Keunecke (1971b, 1972)	Norwegian Coast
Krauss (1959, 1961)	61°N, 22°W
Lisitzin (1953)	Northern Baltic
Magaard and Krauss (1967)	Icelandic waters
Sabinin and Shulepov (1965)	Norwegian Sea
Schule (1952)	Irminger Sea
Zubov (1932)	Barents Sea
Area III	
Darbyshire (1970)	51°N, 7°W
Defant and Schubert (1934)	Baltic
Demoll (1922)	Walchensee
Derügin (1933)	Baltic
Exner (1908a, 1908b)	Wolfgangsee
Exner (1928)	Lunzersee
Gade (1970)	Oslofjord
Gieskes and Grasshoff (1969)	Gotland Basin, Baltic
Gustafson and Kullenberg (1936)	Baltic
Hecht and Hughes (1971)	Bay of Biscay
Hela and Krauss (1959)	Southern Baltic
Hollan (1966b)	54°N, 10°E

TABLE 1-1 (continued)

Reference	Approximate Location
Area III (continued)	
Hollan (1969)	Gotland Basin, Baltic
Johannessen (1968)	Oslofjord
Kielman *et al.* (1969)	Baltic
Krauss (1963)	Baltic
Krauss and Magaard (1961)	Baltic
Kullenberg (1935)	Baltic
Le Floc'h (1970)	Bay of Biscay and Gulf of Guinea
Mortimer (1952, 1953)	Lake Windermere
Pettersson (1916)	57°N, 11°E
Schott (1970, 1971a, 1971b)	56°N, 1°E
Selitskaia (1957)	North Sea
Thorade (1928)	North Sea
Thorpe (1971a)	Loch Ness
Thorpe *et al.* (1972)	Loch Ness
Tomczak (1969)	Western Baltic
Watson (1904)	Loch Ness
Wedderburn (1909b, 1910)	Loch Garry
Wedderburn (1912)	Loch Earn
Wedderburn and Watson (1909)	Loch Ness
Wedderburn and Williams (1911)	Madüsee
Weston and Reay (1969)	51°N, 6°W
White (1967)	52°N, 19°W; 29°N, 18°W
Yampolskii (1962)	55°N, 18°W
Area IV	
Allan (1966)	Gibraltar
Bockel (1962)	Gibraltar
Caloi and Migani (1964)	Lake Bracciano, Italy

TABLE 1-1 (continued)

Reference	Approximate Location
Area IV (continued)	
Caloi *et al.* (1961)	Lake Bracciano, Italy
Cavanie (1972)	Gibraltar
Charnock (1965)	West of Gibraltar
Defant (1940b)	Straits of Messina
Ekman (1953)	West of Spain
Frassetto (1960)	Gibraltar
Glinskii (1960)	Black Sea
Hecht and White (1968)	35°N, 21°W
Helland-Hansen and Nansen (1926)	West of Spain
Horn *et al.* (1971)	30°N, 28°W
Ivanov *et al.* (1969)	Black Sea
Jacobsen and Thomsen (1934)	Gibraltar
Krauss and Düing (1963)	Gulf of Naples
Lacombe (1965)	Gibraltar
LaFond and Rao (1954)	Bay of Bengal
Meincke (1971a, 1971b)	30°N, 28°W
Perkins (1972)	Mediterranean
Shlyamin (1957)	68°S, 15°E
Shonting *et al.* (1972)	Mediterranean
Siedler (1968)	Strait of Bab el Mandeb
Valdez (1960)	10°S, 15°W
Woods and Fosberry (1967)	Mediterranean
Yampolskii (1960)	Black Sea
Ziegenbein (1969, 1970)	Gibraltar
Area V	
Belyakov and Belyakova (1963)	Sargasso Sea
Boston (1963, 1964)	Florida Coast
Brown *et al.* (1955)	Near Bermuda

TABLE 1-1 (continued)

Reference	Approximate Location
Area V (continued)	
Clark and Reiniger (1973)	39°N, 50°W
Cornell Aeronautical Lab (1969)	Cayuga Lake
Csanady (1972)	Lake Ontario
Defant (1940a)	45°N, 39°W
Dowling (1966)	Florida Coast
El-Sabh *et al.* (1971)	Gulf of St. Lawrence
Gaul (1961a, 1961b)	New York Coast
Hale (1965, 1969)	Lake Huron
Halpern (1971a, 1971b)	Massachusetts Bay
Haurwitz *et al.* (1959)	Near Bermuda
Hunkins and Fliegel (1973)	Seneca Lake
Lobb and Hamilton (1969)	Hudson Canyon to Bermuda
Magaard and McKee (1973)	WHOI Site "D"*
Magnitzky and French (1960)	Bahama waters
Malone (1968)	Lake Michigan
Mortimer (1968)	Lake Michigan
Mortimer (1971)	Lakes Michigan and Ontario
Petrie (1973)	Nova Scotian waters
Piip (1969)	Sargasso Sea
Pochapsky (1963)	Puerto Rican waters and the Sargasso Sea
Pochapsky and Malone (1972)	10°N, 50°W
Pollard (1972)	WHOI Site "D"
Regal and Wunsch (1973)	WHOI Site "D"
Schubert (1939)	16°N, 46°W 30°N, 43°W
Schule (1952)	44°N, 41°W Labrador Sea
Seiwell (1937, 1939)	Near Bermuda
Seiwell (1942)	26°N, 54°W

*Around 39°N, 70°W

TABLE 1-1 (continued)

Reference	Approximate Location
Area V (continued)	
Siedler (1971)	WHOI Site "D"
Verber (1964)	Lake Michigan
Voorhis (1968)	WHOI Site "D"
Voorhis and Perkins (1966)	34°N, 66°W
Warner (1972)	Nova Scotian waters
Webster (1963)	34°N, 66°W
Webster (1963, 1968)	WHOI Site "D"
Webster and Fofonoff (1967)	Sargasso Sea
Wunsch and Dahlen (1970b)	Near Bermuda
Wunsch and Hendry (1972)	Massachusetts Coast
Area VI	
Arthur (1954, 1960)	California Coast
Belshé (1968)	Kaulakhi Channel, Hawaii
Cairns (1967, 1968)	California Coast
Cairns and LaFond (1966)	California Coast
Carsola (1967a, 1967b)	California Coast
Carsola and Callaway (1962)	California Coast
Carsola and Jeffress (1968)	Southwest of California
Carsola *et al.* (1965)	Southwest of California
Cox (1960)	California Coast
Düing (1969)	Pokai Bay, Oahu, Hawaii
Emery (1956)	Southwest of California
Hughes (1969)	Georgia Strait
Keunecke (1971a)	California Coast
Knauss (1962)	28°N, 139°W 28°N, 118°W 23°N, 134°W
Krauss (1966a)	Southwest of California
LaFond (1959)	California Coast

TABLE 1-1 (continued)

Reference	Approximate Location
Area VI (continued)	
LaFond (1963)	Around Baja California
LaFond and LaFond (1967)	Around Baja California
LaFond and Moore (1962)	San Diego to Hawaii
Larsen (1969a)	9°N, 88°W
Lee (1961)	California Coast
Lee and Cox (1966)	Southwest of California
Loucks (1963)	50°N, 145°W
Munk (1941)	Gulf of California
Reid (1956)	West of California
Reid (1962)	30°N, 125°W
Rooth and Düing (1971)	Hawaiian waters
Rudnick and Cochrane (1951)	30°N, 140°W 49°N, 148°W
Schule (1952)	40°N, 150°W
Shand (1953)	Georgia Strait
Summers and Emery (1963)	California Coast
Ufford (1945, 1946, 1947)	California Coast
Zalkan (1970)	West of California
Area VII	
Kanari (1968, 1970a, 1970b)	Lake Biwa, Japan
Konaga (1961, 1965)	Japanese Coast
Lek (1938)	11°S, 121°E
Lek and Fjeldstad (1938)	1°S, 126°E
Maeda (1971)	29°N, 135°E
Nan'niti and Yasui (1957)	27°N, 140°E
Perry and Schimke (1965)	Andaman Sea
Tominaga (1970)	31°N, 127°E 30°N, 142°E

TABLE 1-2

Summary of Observations for Each Map Area

	Number of observations	Maximum vertical displacement*	Periods*	Lengths*	Maximum current speeds*
Short-period					
Area I	4	2-10 m	10 min		
Area II	1		7-13 hr		
Area III	5	10 m	9 min-6 hr	5×10^3 m	0.5 m/sec
Area IV	7	0.5-40 m	5 min-5 hr	10^3 m	
Area V	3	10 m	3-15 min		
Area VI	8	1-12 m	3 min-8 hr	10^2-1.2×10^3 m	0.25 m/sec
Area VII	3	82 m	10 min-4 hr	2×10^3 m	
Lakes	3		5 min-6 hr	10^3 m	
Semi-diurnal					
Area I	1				
Area II	6	10-64 m		2.2×10^4-3×10^4 m	0.1 m/sec

Area III	8	2-15 m	4.5×10^4 m	0.4 m/sec
Area IV	11	60-180 m		
Area V	14	7-50 m	10^5 m	0.17-0.5 m/sec
Area VI	13	10-150 m	9×10^3-1.7×10^5 m	0.2 m/sec
Area VII	9	14-70 m	1.3×10^5 m	
Seiches				
Lakes	12	10-23 m		0.25 m/sec
Diurnal				
Area II	2	77 m		
Area III	1			
Area IV	3			
Area V	4	10-100 m		0.08 m/sec
Area VI	3			
Area VII	4	15-30 m		

TABLE 1-2 (continued)

	Number of observations	Lengths*	Horizontally coherent?	Vertically coherent?	Maximum current speeds*
Inertial					
Area I	2		yes		
Area II	1				
Area III	6		no	yes	0.06-0.5 m/sec
Area IV	4		yes	yes and no	0.1 m/sec
Area V	6		yes	no	0.6 m/sec
Area VI	2	10^3- 2×10^6 m		no	
Area VII	1				

* Values are ranges observed.

already been attempted for the wavenumber-frequency energy spectrum (Garrett and Munk, 1972b).

Table 1-3 gives a summary of observed wavelengths for a wave of a given period. One is struck with the paucity of experimental information for such a basic wave relationship.

Hinwood (1972) has recently given an excellent summary of observations of internal waves up to 1945.

1.2 FIELD INSTRUMENTATION

While internal waves were known to exist in the ocean in the early 1900s (Ekman, 1904), it is only within the last two decades or so that technology has advanced to a point where the large number of observations required for their statistical description are available. Even so, the problems connected with measuring motion within the ocean are enormous. This section gives a brief summary of some of the methods currently being developed and in use for such measurements. A review of equipment used until about 1960 to measure internal waves may be found in LaFond (1962). Krauss (1966b, pp. 147-148) also has given a review of internal wave instrumentation.

Most of the recent internal wave measurement methods have been developed with the idea of preventing falsification (or "contamination") of internal wave data by temperature and salinity microstructure. There are several approaches to this problem.

1. Ignore it. This is becoming less and less acceptable.
2. Try to sample where the temperature or salinity profile is known to be smooth. However, because of the precision of today's instruments, a smooth profile in the ocean is hard to come by.
3. Use a reliable instrument to gather data at a fixed point and correct for contamination in the analysis. This is the approach that Webster and Hasselmann from Woods Hole are using in their ambitious project to collect a long series of data in the deep Atlantic (Webster 1972: personal communication). The project, which is scheduled to start in 1973, involves mooring an array of current meters and thermometers in the Atlantic Ocean where the depth is 5.5 km and taking temperature and current measurements in the main thermocline. Correction

TABLE 1-3

Observed T-λ Relationship

Wave period	Wavelength (m)	Reference	Approximate location
2.5 min	54	Hughes (1969)	Georgia Strait
5 min	10	Woods and Fosberry (1967)	Mediterranean
8 min	100	Cairns (1967)	California Coast
10 min	700	Ziegenbein (1969)	Gibraltar
10-15 min	300	Larsen (1969a)	Costa Rican Coast
13 min	10^3	Ziegenbein (1970)	Gibraltar
15 min-3 hr	400	Ufford (1946)	California Coast
50 min	300	Byshev *et al.* (1971)	Black Sea
80 min	400	Byshev *et al.* (1971)	Black Sea
3 hr	5×10^3	Schott (1971a)	North Sea
3.5 hr	1.1×10^3	Byshev *et al.* (1971)	Black Sea
4.5 hr	9×10^3-1.2×10^4	Keunecke (1971a)	California Coast
5 hr	1.6×10^3	Byshev *et al.* (1971)	Black Sea
Around 12 hr	9×10^3	Cairns (1967)	California Coast
	2.2×10^4	Keunecke (1971b)	Norwegian Coast
	2.5×10^4	Keunecke (1972)	Norwegian Coast
	3×10^4	Magaard and Krauss (1967)	Faeroe Ridge
	4.5×10^4	Schott (1971a)	North Sea
	10^5	Seiwell (1942)	Mid-Atlantic
	1.7×10^5	Summers and Emery (1963)	California Coast
Inertial	5×10^4	Schott (1971b)	North Sea
	4×10^4-2×10^6	Pollard (1972)	Massachussetts Coast

for microstructure contamination of data from the moored array has been investigated by Joyce (1973).

4. Have a sensor move through the water column and take profiles of temperature, salinity, and/or currents as often as possible, then reconstruct the isotherms or isopycnals. This general approach was used when BT's first came out, but was by and large abandoned when thermistors made it easier to take temperature measurements at a fixed depth. It has lately come back into vogue, at a level of sophistication ranging from a single thermistor (Neshyba, Neal, and Denner, 1972) to the cycling profiling current meter being developed at RSMAS, University of Miami (Van Leer *et al.*, 1974). A prototype of the latter has been described by Düing and Johnson (1972), but the meter now being developed samples temperature, conductivity, and pressure, as well as horizontal current, continuously through the water column. Scripps Oceanographic Institution is developing a constant mass system which is designed to operate at depths of around -10^3 m. It consists of a floating capsule which can flow horizontally with water of constant density; at the same time, it moves up and down 20 m every 12 min taking temperature profiles (Cairns 1973: personal communication). One of the recent uses of FLIP, the floating instrument platform of Scripps, is to provide a stable platform while thermometers are lowered rapidly through the water column and obtain many temperature profiles (Munk 1972: personal communication). The STD is sometimes used for internal wave measurements (Morse and Smith, 1972); it is unsuitable, however, for looking at microstructure (Osborn 1972: personal communication). The response time of an STD thermometer, for one thing, can be a problem (Goulet and Culverhouse, 1972).

5. Inject dye into the water and observe its behavior (Woods, 1968; Van Leer, 1971). This technique has obvious disadvantages, but the simplicity of the method should not lead one to ignore its advantages.

There have been several attempts to use floats to measure the movement of isopycnals during passage of an internal wave, but the method is fraught with problems. Pochapsky (1963) has shown that even in the upper waters of the ocean, one needs to be wary when attempting to correlate temperature fluctuations with water motion. A recently developed pycnocline follower (Düing, 1970) overshoots internal waves and gives amplitudes as much as twice as large (Düing 1972: personal communication). The work on the pycnocline follower has led to development of the cycling profiling current meter, mentioned above. Proni and Apel (1973) have observed internal waves using a high-frequency transducer.

This section would be incomplete without mentioning the large moored array of Wunsch and Dahlen (1970b) near Bermuda, and the observational masts in the Baltic used by the Institut für Meereskunde, University of Kiel, Germany (Krauss, 1966b, p. 148).

1.3 GENERATION OF INTERNAL WAVES

There are few unequivocal observations of internal waves being generated in the ocean. For one thing, dissipation rates for oceanic internal waves are not known, so it is difficult to say with any degree of certainty that internal waves observed at point B were generated at point A. Nevertheless, there are instances where a proposed generating mechanism for observed internal waves is plausible.

Changes in topography, e.g., slopes, sills, seamounts, may produce internal waves. One of the earliest observations was made by Reid (1956). He used hourly BT data taken from three ships in a line perpendicular to the coast of California and at distances from the coast of 7.4×10^4 m (depth = -2160 m), 2.7×10^5 m (depth = -4700 m), and 5.9×10^5 m (depth = -4830 m). The amplitudes of the semi-diurnal internal tide were greatest (about 10 m) at the station closest to the shore; at the other two stations, the internal tide was almost absent. He concluded that the internal tide was probably being generated at the shelf. From recent experiments, Regal and Wunsch (1973) and Wunsch and Hendry (1972) have concluded that the semi-diurnal internal tide off the Massachusetts Coast is probably being generated on the continental slope in the region where the slope becomes critical for the tidal period (see Sec. 4.4).

It is possible that internal waves may be generated somehow by irregularities in the ocean floor (Cox and Sandstrom, 1962). Zalkan (1970) has proposed that the short-period internal waves which he observed off the California Coast might be generated by seamounts about 1.2 × 10^5 m away, while Schott (1971a) has suggested that the internal tide in the North Sea might be generated in the Dogger Bank region about 2 × 10^5 m away from the site of his observation. The Mid-Atlantic Ridge (Sabinin and Shulepov, 1970), the Izu-Mariana Ridge in the Pacific (Maeda, 1971), and the Straits of Gibraltar (Gade and Ericksen, 1969) have also been claimed as generating sites for internal waves.

There are reported instances of short-period internal waves being somehow generated by long-period ones: Gade and Ericksen (1969) have suggested that short-period (10 to 40 min) internal waves east of the Straits of Gibraltar are possibly stable lee-waves formed in the trough of the internal tide. Hecht and Hughes (1971) have observed that short-period internal waves are somehow related to the internal tide in the Bay of Biscay, and Cairns (1967) has found the same result on the California Coast. Halpern (1971a) and Lee (1961) have shown that short-period internal waves are associated with the change of the tide. From a specialist's point of view, the work of Gargett (1970, cited by LeBlond 1972: personal communication) in Georgia Strait showing the phase relationship of trains of internal waves and the tides is very important.

Internal waves may be generated when a ship moves slowly through highly stratified water, the so-called "dead water" effect (Ekman, 1904). Hughes at the Department of the Environment in Victoria, B. C., Canada, is presently working on a problem which involves a slowly moving ship creating internal waves. He has some good aerial photographs of the resulting surface patterns (see Sec. 4.10).

It has long been known that winds cause internal seiches in lakes (Exner, 1908a; Mortimer, 1953; Kanari, 1970a). Mortimer (1971) has given a very good account of the internal waves and seiches in Lake Michigan and their generation by wind. Recently, Thorpe, Hall, and Crofts (1972) have done a definitive experiment in Loch Ness showing how the internal seiche is created: Wind stress on the water surface causes a flow of near-surface water to one end of the loch and a depression of the thermocline at that end. When the wind falls, the tendency of the thermocline to recover a level position initiates the seiche. They have shown that the seiche develops a pronounced front or surge like a tidal bore; this is followed by a periodic oscillation of the isotherms. A similar shape for temperature records has been noted by Hunkins and Fliegel (1973) in Seneca Lake, New York. In their case, however, there was no simple correlation between the surges and changes in either air pressure or wind speed and direction. The surge is often over 10 m in height and the following wavetrain, when it occurs, is of comparable height and contains waves about 10^3 m long. Recordings have been made at several fixed locations showing the repeated passing of the surge as it progresses up and down Loch Ness for well over a week, during which time it has propagated for a distance of over 2×10^5 m. During this time there is no visible effect at the surface of the loch. Thorpe *et al.* (1972) have cited the following example: On the evening of 30 September 1971, a strong southwest wind blew up (Loch Ness is oriented S.W.-N.E.) and continued until late in the afternoon of 1 October. Fig. 1-2 shows the results of temperature and current measurements taken in the middle of the loch at a depth of 40 m, starting on 2 October.

It is likely that winds, air-pressure fluctuations, and surface swell are capable of creating internal waves in the ocean. Schott (1971b) has observed an instance in the North Sea when wind speeds of 15 m/sec seemed to create internal oscillations with periods below the inertial period. Krauss (1959) has found internal waves of inertial period in the northeast Atlantic which seem to have been caused by air-pressure fluctuations, and Pollard (1972) has also found that inertial waves in the Atlantic may be generated by storms. An interesting case of possible generation of short-period internal waves (period 10-15 min) by a storm has been reported by Larsen (1969a). Krauss (1967) has proposed that short-period internal waves on the California Coast are resonantly generated by swell (see also Sec. 2.2), and the measurements he made seemed to confirm the hypothesis.

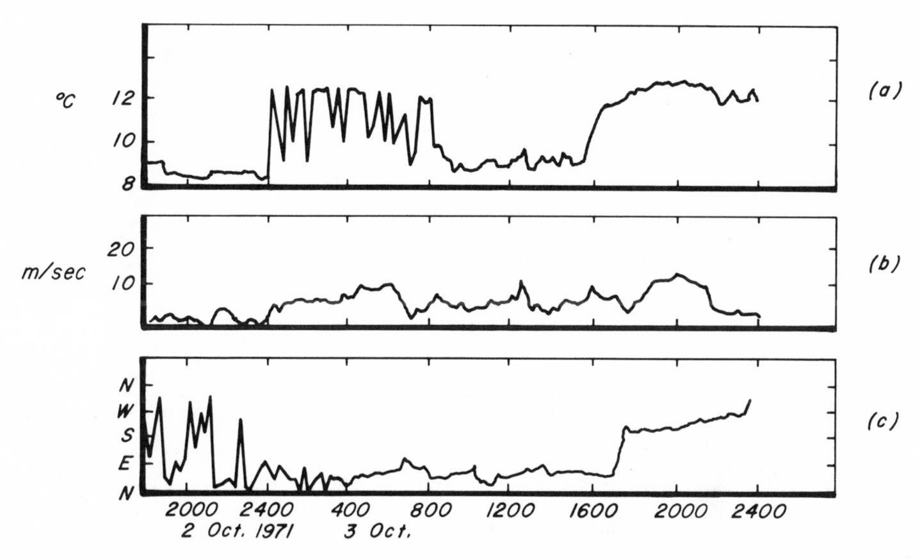

FIG. 1-2. (a) Temperature at 40 m in Loch Ness in October 1971; (b) magnitude of the current at 40 m; (c) direction of the current. (Redrawn from Thorpe, Hall, and Crofts, 1972, and used with Dr. Thorpe's permission.)

In the atmosphere, lee waves are observed to be generated when the wind blows over mountain ranges. Miles (1968) has given a summary of some lee-wave generation observations. Malkus and Stern (1953) have observed that lee waves can be created when the wind blows gently over a heated island.

1.4 SHORT-PERIOD INTERNAL WAVES

General remarks

The periods of the shortest internal waves are much greater than those of surface waves of comparable length, and their speeds are correspondingly slower, about the speed of a walk. Their wave slopes (i.e., the ratio $2a:\lambda$) are quite small. One of the most remarkable things about internal waves is how little they affect the water's surface. Larsen (1969a) has told of measuring, aboard the R. V. *Thompson*, internal waves only 20 m below the surface which had vertical displacements as great as 5 m yet, as he said, "One could have played pool aboard the *Thompson*." Short-period internal waves are a transient phenomenon and, when they occur, tend to occur in packets. In the thermocline in shallow water, internal waves have broad crests and steep troughs.

Period

For the purposes of this section, "short-period internal waves" are those whose periods are significantly less than 12 hr. Their periods can be as short as 2.5 min (Hughes, 1969; in Georgia Strait) or as long as 5 hr (Ivanov *et al.*, 1969; in the Black Sea), but most have periods ranging between 5 and 20 min. Off the coast of California, there seems to be a persistent peak in the energy spectrum for periods around 18 min. Carsola (1967a) found it in 1965 using a moored thermistor, and Pinkel at Scripps Oceanographic Institution (1972: personal communication) found it in 1971 using data from frequent temperature profiles.

Amplitude

Observed amplitudes for short-period internal waves vary from 0.2 m in the Mediterranean (Woods and Fosberry, 1967) to 40 m in the Andaman Sea (Perry and Schimke, 1965). Typical amplitudes of several meters have been seen in the Arctic Ocean (Neshyba, Neal, and Denner, 1972; Morse and Smith, 1972); amplitudes of 3 or 4 m have been measured off the coast of California (LaFond, 1962; Carsola and Callaway, 1962) and in Massachusetts Bay (Halpern, 1971a). Short-period internal waves near the Straits of Gibraltar have amplitudes of up to 20 m (Ziegenbein, 1969) and, as noted above, Perry and Schimke have reported internal waves in the Andaman Sea with periods of about 20 min whose amplitudes were over 40 m.

Wave packets

One of the most noted characteristics of short-period internal waves is their tendency to occur intermittently in packets. These packets are sometimes associated with long internal waves (Halpern, 1971a; Hecht and Hughes, 1971; Lee, 1961; Shand, 1953; Ziegenbein, 1969; Woods and Fosberry, 1967). In other instances, no such relation is indicated (Hughes, 1969; Gaul, 1961a; Tominaga, 1970; Ivanov *et al.*, 1969; Zalkan, 1970).

Wavelength

It is next to impossible to get a picture of usual wavelengths for short internal waves. Garrett and Munk (1972b) have pointed out this difficulty in trying to construct dispersion relations for internal gravity waves. By photographing the trace of dye injected into the thermocline in the Mediterranean, Woods and Fosberry (1967) have observed wavelengths of 10 m for waves with periods of 5 min; the amplitudes were about 0.2 m. This wavelength seems to be quite a bit shorter than is usually reported. For instance, the internal waves observed by LaFond (1963) off Baja California range in length from 100 to 1.5×10^3 m; however, the period of these waves is not specified. In any case, short-period internal waves have small slopes; even the relatively steep waves observed by Woods and Fosberry have a ratio $2a:\lambda = 1:25$.

Crest length

The crest length of waves with periods of about 10 min may be inferred to be around 8×10^3 m: Halpern (1971a) has reported seeing surface bands, somehow associated with internal waves in Massachusetts Bay, of length 9.7×10^3 m, and Ziegenbein (1969) has reported bands of length 7.4×10^3 on the Mediterranean.

Phase speed

The phase speed of short-period internal waves varies between 0.1 m/sec (LaFond, 1962; Ivanov *et al.*, 1969), and 1 m/sec (Halpern, 1971a; Ziegenbein, 1969). Caution must be exercised in juggling these numbers back and forth using $\lambda = c_p T$. Many of the authors have already made this tacit assumption in reporting their observations, and one may end up by concluding a premise.

Direction

Near the coast, short-period internal waves seem to move shoreward (Gaul, 1961a; LaFond, 1962). Otherwise, no general conclusions may be drawn about their direction of propagation.

Currents

The currents associated with short-period internal waves may be appreciable: Hughes (1969) has found currents as great as 0.25 m/sec associated with waves of 2.5-min periods in Georgia Strait. In contrast, LeFloc'h (1970) has found current speeds of only 0.05 m/sec associated with waves with periods of 25 min in the Bay of Biscay.

Energy

Voorhis (1968) has measured horizontal and vertical currents off the Massachusetts Coast. A comparison of the spectra for horizontal velocity and vertical velocity (and thus for kinetic energy), and potential energy shows that below periods of about 10 hr and above the Brunt-Väisälä period T_N (which is about 1 hr at the depth Voorhis considered), potential energy and kinetic energy are approximately equal. (For periods greater than 10 hr and less than 1 hr, potential energy is less than kinetic energy). The ratio of the horizontal component of kinetic energy to the vertical component is always greater than one. The values of this ratio for periods between 10 and 1 hr have led Voorhis to suggest that the particle trajectories in this period range are elliptical, with the major axis horizontal and the minor axis vertical. For periods around 10 hr, the ratio of major to minor axis is about 10:1, while around 1 hr, it is about 3:2.

Waveform

It was LaFond (1961) who first observed that internal waves on the thermocline are not sinusoidal but are flattened on the crests when the thermocline is shallow and peaked on the crests when the thermocline is deep. The general shape of a progressive internal wave on a shallow thermocline is shown in Fig. 1-3.

1.5 LONG-PERIOD INTERNAL WAVES

In this section are included internal waves with periods longer than about 12 hr. Internal waves with periods anywhere around 12 hr

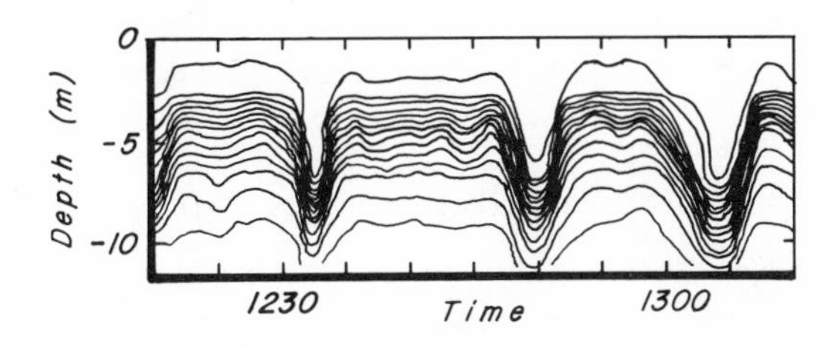

FIG. 1-3. Isotherms off Southern California, 12 June 1958. Water depth is about 20 m. (Redrawn from LaFond, 1962.)

are called *semi-diurnal internal tides*, and internal waves with periods anywhere around 24 hr are called *diurnal internal tides*, even though the term "tide" may be a misnomer. In the literature, internal waves with these periods are also referred to as *baroclinic tides*.

General remarks

In general, internal tides have periods only approximately equal to the periods of surface tides; have wavelengths much greater than their amplitudes; travel with speeds of around 0.5 m/sec; and have amplitudes which are affected by coastal topography. They seem to occur more consistently than do internal waves with shorter periods. Nonetheless, they are not always present in the ocean, especially not in the open ocean (Haurwitz, Stommel, and Munk, 1959; Rudnick and Cochrane, 1951; Magaard and Krauss, 1967; Krauss, 1966a; Hecht and Hughes, 1971; Reid, 1962). In addition, many claimed observations of internal tides are open to question because of insufficient data (Haurwitz, 1954). Hendershott (1973) has reviewed work on ocean tides, including internal tides.

Period

While the surface tide is very regular (a Fourier transform of a surface-elevation record almost invariably gives amplitude maxima precisely at the tidal periods) internal wave records give amplitude maxima at periods near, but significantly different from, the tidal periods. A likely explanation is that internal waves, whose scales are small compared to those of the surface tide, are being Doppler-shifted by ocean currents (Regal and Wunsch, 1973).

Amplitude and wavelength

The amplitudes of internal tides, usually 2-10 m, are very small compared to their wavelengths, which are usually around 3×10^4 m. For instance, Schott (1971a) has found that the semi-diurnal internal tide in the North Sea has an amplitude of 2 m and a wavelength of 4.5×10^4 m, giving a ratio of $2a:\lambda \doteq 1:(10^4)$. Even with much larger amplitudes the wave slope is small; for instance, near the Iceland-Faeroe Ridge, Maggard and Krauss (1967) have found semi-diurnal amplitudes as great as 50 m, with wavelengths about 3×10^4 m, giving a ratio of $2a:\lambda = 1:(3 \times 10^2)$. Very large semi-diurnal vertical displacements are sometimes indicated; Bockel (1962) has observed vertical displacements of 180 m in the Straits of Gibraltar (the wavelengths are not known).

In Massachusetts Bay, Halpern (1971b) has found the following amplitudes for the first five modes of the semi-diurnal internal tide: first mode, 3.676 m; second mode, 0.879 m; third mode, 1.707 m; fourth mode, 2.699 m; fifth mode, 2.220 m. The low amplitude of the second mode internal wave is unexplained (Halpern 1972: personal communication).

Direction

There is indication that internal tides propagate shoreward on the Norwegian and California Coasts (Keunecke, 1971b and 1972; LaFond, 1962). Using simultaneous data for three 24-hr periods taken by six to ten ships off the California Coast, Summers and Emery (1963) have constructed a map of the progress of the crest of one such wave (Fig. 1-4). In other parts of the ocean, the direction of the semi-diurnal tide may vary. Along the western Florida Coast, Boston (1964) has found that the semi-diurnal tide proceeds in the same direction as the surface tide, while in the North Sea, Schott (1971a) has found that it propagates at right angles to the surface tide.

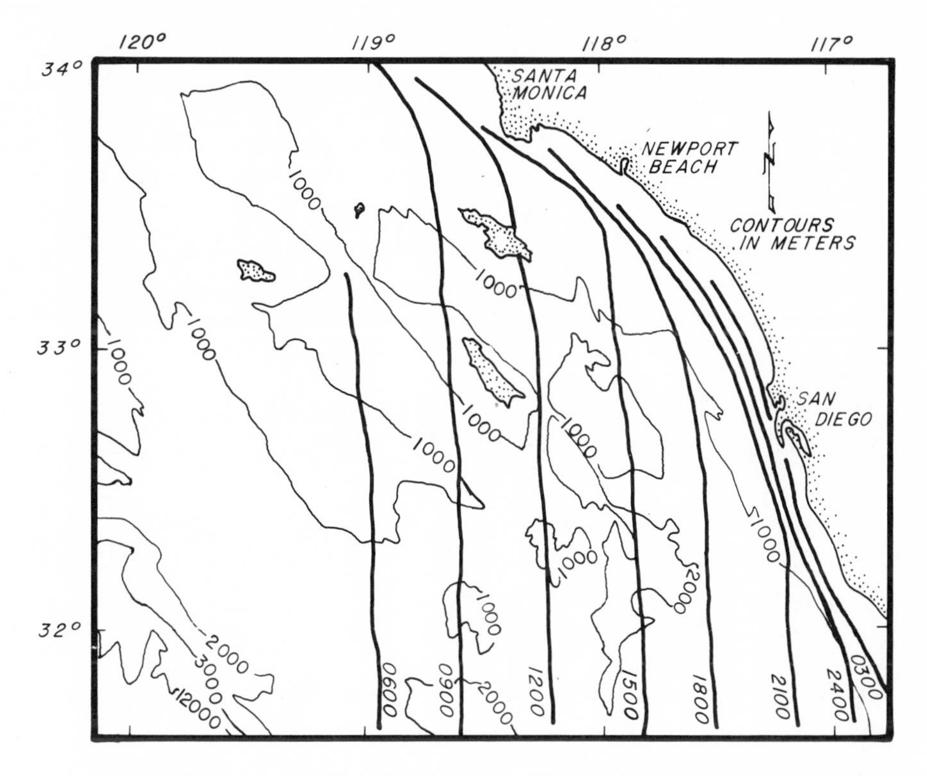

FIG. 1-4. Positions of the crest of the semi-diurnal internal tide from 0600 on 19 August to 0300 on 20 August 1961, constructed from BT data taken from a multiple-ship survey. (Redrawn from Summers and Emery, 1963.)

Amplitude and topography

Near the coasts, the amplitude of the semi-diurnal tide is influenced by bottom topography. On the continental slope of Norway, Keunecke (1971b and 1972) has found that the amplitude is less than 5 m outside the shelf region, about 25 m at the slope, and about 10 m on the shelf. The wavelengths are around 2.5×10^4 m. Similar results have been found by Reid (1956) and Carsola (1967b) off the California Coast. In analyzing data from a thermistor chain towed from California to Hawaii, Krauss (1966a) has found semi-diurnal internal tides near California, but none beyond the halfway point. Sec. 1.8 discusses further observed topographic effects.

Currents

Currents associated with the semi-diurnal internal tide vary from around 0.1 m/sec off the Norwegian Coast (Fjeldstad, 1964) to over 0.4 m/sec in the North Sea (Schott, 1971a). Currents of 0.2 m/sec seem to be average (Reid, 1958; Boston, 1964; Meincke, 1971b; Defant, 1940a; Halpern, 1971b).

Phase

Phase relationships between the surface and internal tides appear to be variable. Off the California Coast, the internal tide lags the surface tide by 3 to 6 hr (Cairns and LaFond, 1966; Carsola, 1967b); off the Massachusetts Coast, the two seem to be in phase (Regal and Wunsch, 1973); and on the continental slope off Nova Scotia, there appears to be a 180° phase shift (Petrie, 1973).

Seasonal variability

Taking seasonal hydrographic data for the North Atlantic and using numerical methods to obtain a description of internal wave eigenfunctions and currents (Sec. 3.4), Boris (1970a, 1972) has constructed maps which describe various aspects of internal tidal motion, including maximal amplitudes and currents, by season.

Diurnal internal tide

Evidence for the diurnal internal tide is found less often in internal wave data than for the semi-diurnal internal tide. Theory predicts it is not possible for the diurnal internal tide to exist as a free wave north of 30°N (Phillips, 1966, p. 196). Nonetheless, internal waves with periods of around 24 hr have been reported considerably north of that latitude. In the Irminger Sea, for instance, Krauss (1961) has found the vertical displacement at 500 m for internal waves of 24-hr period to be as great as 77 m, compared to 64 m for internal waves of 12.5-hr period. Off the California Coast, the semi-diurnal internal tide predominates, although as one goes out from shore the amplitudes of the two become comparable (Krauss,

1966a). Near the Great Meteor Seamount in the Atlantic (30°N, 28°W), currents associated with the semi-diurnal internal tide are as great as 0.2 m/sec, while those for the diurnal internal tide are around 0.1 m/sec (Meincke, 1971b).

Energy

In the Atlantic, the kinetic and potential energies are approximately equal for internal waves with periods around 10 hr; for waves with longer periods, potential energy is less than kinetic energy (Voorhis, 1968). There is indication that, in the Atlantic at least, horizontal kinetic internal wave energy propagates vertically (Frankignoul and Strait, 1973).

Inertial waves

Internal waves with the longest periods are *inertial waves*, which for our purposes are defined as waves whose periods are very close to the inertial period for a particular latitude. While the vertical displacements associated with these waves may be large enough to be measured (Schott, 1971b; Rooth and Düing, 1971; Pochapsky, 1966; Shonting *et al.*, 1972), they are negligible compared to their wavelengths, which are on the order of hundreds of kilometers (Pollard, 1972). Besides the fact that observed inertial waves may have vertical displacements, while true inertial waves do not, the observed period for inertial waves seems to be 1% to 3% less than the true inertial period (Perkins, 1972, 1973; Schott, 1971b; Kielmann, Krauss, and Magaard, 1969).

The currents associated with inertial waves can be quite strong, from 0.5 to 0.6 m/sec, and at times can dominate the mean current. Schott (1971b) has discovered that inertial currents in the North Sea go in opposite directions on opposite sides of the thermocline.

As with other internal waves, inertial waves are a transient phenomenon, often persisting for only a few cycles. It has been suggested (Day and Webster, 1965) that this may be due to their generation by winds. They have been found at many locations in the northern hemisphere (Webster, 1968) and in the deep ocean (Pochapsky, 1966).

Besides using neutrally buoyant floats in attempting to get a picture of inertial oscillations, investigators have used current meters set in moored vertical and/or horizontal arrays, and then examined the vertical and/or horizontal coherence. No doubt the resulting confusion is in part due to the fact that, as Schott (1971b) has pointed out, stratification of the water and precise determination of current-meter depth are crucial factors. In a recent paper, McWilliams (1972) has pointed out that in some cases the vertical correlation of energy may be quite good, even though the phases show little vertical coherence. Results of different investigations are difficult to compare because of the varying sizes of the arrays, to say nothing of their varying locations. A few of these results are summarized in Table 1-4.

Webster (1968 and 1970) has given two excellent summaries of inertial wave observations.

Internal seiches

Standing internal waves, or *internal seiches*, in lakes gave one of the earliest evidences for the natural occurrence of internal waves. In 1903, Sir John Murray, E. R. Watson, and E. M. Wedderburn observed

TABLE 1-4

Horizontal and Vertical Coherence of Inertial Waves at Various Locations

Location	Reference	Vertical coherence	Horizontal coherence
Site "D"	Webster (1968)	low	high
Baltic	Kielmann *et al.* (1969)	high	low
Upper Mediterranean	Shonting *et al.* (1972)	low	high
Deeper Mediterranean	Perkins (1972)	high	
Arctic Ocean	Bernstein and Hunkins (1971)		high

that the isotherms of Loch Ness were "continually and quickly altering their inclinations." They also observed that "the area between two consecutive isotherms remained practically constant from one day to the next, although of course the shape of this area had entirely altered" (Watson, 1904). A plot of the temperature fluctuations vs. time showed a definite periodicity, which Watson explained as an internal seiche, using the analogy of oil over water in a glass tank. Watson's conclusions were questioned later by Birge (1909), who suggested that the large displacements reported by Watson, over 30 m, were temperature variations caused by the wind. Birge wrote, "If I may hazard a doubtful opinion, I would say that I believe such seiches may be found to exist. I believe also that in most lakes they will be of small extent and of small influence upon the temperature." Birge, by the way, introduced the term "thermocline" into the literature in 1897, and his paper in 1909 introduced two more new terms: "epilimnion" for the upper warm layer of water in a lake, and "hypolimnion" for the lower, colder water.

Observations of internal seiches, whose frequency is dependent upon the shape of the enclosing basin, have been reported in many different locations (Table 1-5). Internal seiches in the Baltic have been extensively studied, and a summary of the work has been given by Krauss (1969). These seiches have periods of 39, 22.5, and 13 hr; their wavelengths are very short, less than 2×10^3 m; and histograms of currents and temperature fluctuations are not Gaussian - that is, normal distributions are very seldom found.

1.6 SOLITARY INTERNAL WAVES

Solitary internal waves of depression have been observed off the California Coast (Lee, 1961) and in the Mediterranean (Ziegenbein, 1970). Temperature records taken with a triangular array near Hudson Canyon indicate that solitary internal waves of depression (Fig. 1-5) are a common occurrence (Gaul, 1961a). For a considerable time before and after the passage of a single internal depression, little or no temperature activity is recorded; a single wave appears out of a

TABLE 1-5

Observations of Internal Seiches

Location	Reference
Lakes and enclosed seas	
Baltic Sea	Hela and Krauss (1959) Kielmann *et al.* (1969) Krauss (1963, 1969) Krauss and Magaard (1961) Wedderburn (1909a)
Black Sea	Glinskii (1960)
Lake Biwa, Japan	Kanari (1970a)
Lake Cayuga, New York	Cornell Aeronautical Lab (1969)
Lake Michigan	Mortimer (1963)
Lake Seneca	Hunkins and Fliegel (1973)
Lake Windermere	Heaps and Ramsbottom (1966) Mortimer (1952, 1953)
Loch Earn	Wedderburn (1912)
Loch Garry	Wedderburn (1910)
Loch Ness	Thorpe (1971a) Thorpe *et al.* (1972) Watson (1904) Wedderburn and Watson (1909)
Lunzersee	Exner (1928)
Madüsee, Austria	Halbfass (1911) Wedderburn and Williams (1911)
Mondsee	Halbfass (1909)
Walchensee	Demoll (1922)
Basins, gulfs, and straits	
Catalina Basin, California Coast	Emery (1956)
Florida Straits	Mooers (unpublished) Niiler (1968)
Georgia Strait	Gargett (1968, 1970)
Gulf of California	Munk (1941)
Strait of Bab el Mandeb, Red Sea	Siedler (1968)

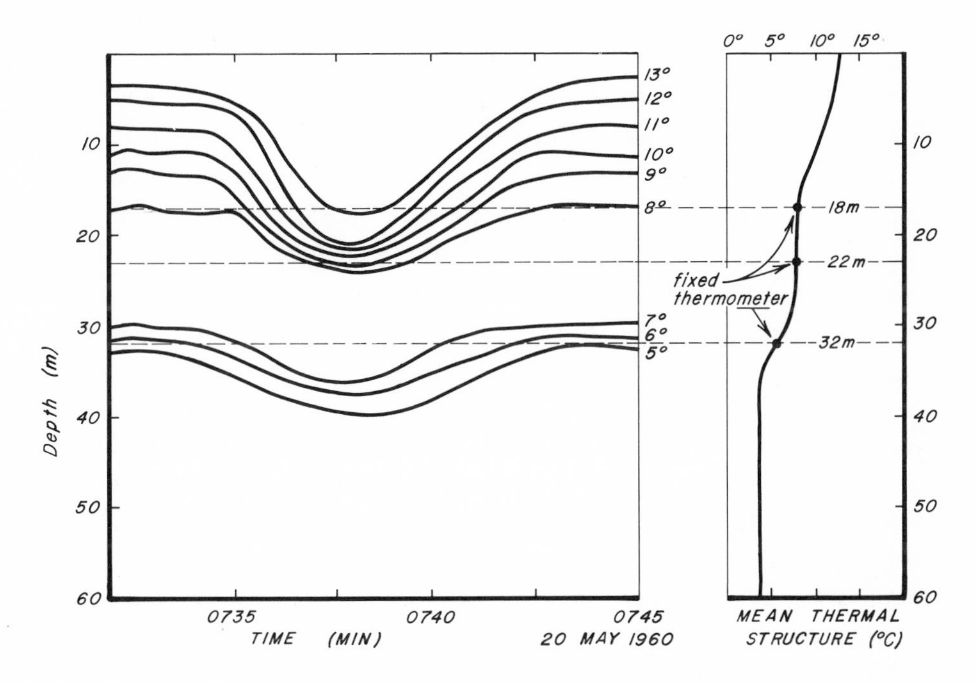

FIG. 1-5. Isotherm depths constructed from temperature data at three fixed depths during passage of a solitary internal wave. The wave is shown for only one element in the triangular array. (Adapted from Gaul, 1961a.)

quiescent background about as frequently as wavetrains appear during periods of marked activity. In analyzing the data of Ivanov *et al.* (1969) taken in the Black Sea, Byshev, Ivanov, and Morozov (1971) noticed occasional perturbations behaving like solitary waves of depression. However, the data were insufficient for a statistical description.

1.7 WAVETANK EXPERIMENTS

There have been many elegant and interesting wavetank experiments which illustrate various properties of internal waves; only the most recent ones are mentioned here. Thorpe (1968c) and Miles (1968) have given a review of much of the older work; Thorpe's paper includes an account of his wavetank experiments on the shape of progressive internal waves (see Sec. 3.5).

The patterns, propagation, and amplitudes of internal wave crests have been investigated for a variety of motions: a cylinder oscillating horizontally (Mowbray and Rarity, 1967a); a cylinder moving with constant velocity (Stevenson, 1968); a cylinder oscillating as it moves (Stevenson and Thomas, 1969); a sphere moving steadily vertically (Mowbray and Rarity, 1967b); a sphere moving steadily horizontally (Lofquist, 1970); a sphere oscillating as it moves vertically (Stevenson, 1969); a bobbing sphere and a buoyantly rising fluid (McLaren *et al.*, 1973); a collapsing wake (Wu, 1969; Schooley and Hughes, 1972). In all of these cases, it seems that linearized theory predicts the phase configuration of the waves reasonably well.

The experiments of Mowbray and Rarity (1967a) are of general interest, because they illustrate the fact that the group and phase velocities of sufficiently short internal waves are nearly perpendicular (see Sec. 3.4). In their experiments, a horizontal cylinder or rod is oscillated in simple harmonic motion with frequency ω in a fluid for which the Brunt-Väisälä frequency N = constant. If $\omega < N$, the wave propagates away from the rod, the over-all pattern being that of a St. Andrew's cross (Fig. 1-6). The arms of the cross increase in length at a uniform rate. Within each arm, the crests, or waves of constant phase, move in a direction toward the horizontal. In the shadowgraphs, these lines of constant phase appear as dark (or light) bands forming on one side of the arm, moving across to the other side in the direction of the phase velocity $\vec{c}_p$, and then disappearing. No disturbances are found outside the region of the cross. If the rod oscillates so rapidly that $\omega > N$, there are no waves. The arms of the cross point in the directions of the group velocity. The angle of inclination of the arms to the horizontal, θ, is determined by the stratification and by how fast the rod oscillates; it is given by

$$\sin\theta = \frac{\omega}{N}$$

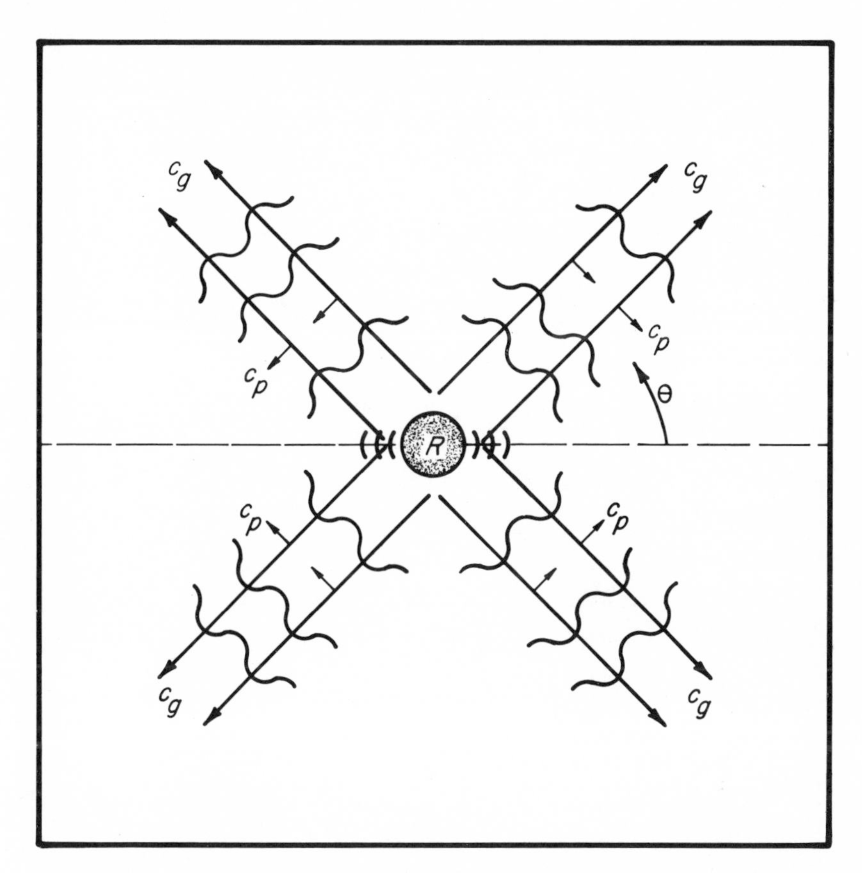

FIG. 1-6. Schematic representation of the wavetank experiment of Mowbray and Rarity (1967a). R is a horizontal cylinder, viewed from the end, which oscillates back and forth. The long arrows point in the direction of $\vec{c}_g$, the group velocity, and their shafts are lines of constant phase; the short arrows point in the direction of $\vec{c}_p$, the phase velocity.

(see Eq. 3.4.25). This experiment has been repeated by Thomas and Stevenson (1972) and by Gordon and Stevenson (1972), who have taken viscosity into account; Gordon and Stevenson also considered what happens when $\omega = N$.

Keunecke (1970) has generated standing internal waves in a wavetank by periodically varying the air pressure at the surface. Photographs of the waveforms and streamlines are given for different pressures. He has found that the wave amplitude is directly proportional to the amplitude of the surface pressure variation. The particle trajectories for internal waves generated by a buoyantly rising fluid have been discussed by McLaren *et al.* (1973).

The wavetank experiments of Joyce (1972), Lee (1972), Martin, Simmons, and Wunsch (1972), Parrish and Niiler (1971), and Weigand *et al.* (1969), are discussed in Secs. 2.2, 2.5, 2.2, 2.4, and 2.4, respectively. Observations of breaking internal waves in wavetanks are discussed in the next section.

1.8 SHOALING AND BREAKING INTERNAL WAVES IN THE OCEAN, IN WAVETANKS, AND IN THE ATMOSPHERE

Incontrovertible evidence that high-frequency internal waves in the ocean actually do break has been provided in the last few years. In the Mediterranean, in the waters off Malta, Woods *et al.* have done a series of experiments in which scuba divers injected colored dye into the thermocline and then photographed the results (Fig. 1-7). Previous studies (Woods and Fosberry, 1967) have shown that the thermocline off Malta is made up of *layers* 2-4 m thick of nearly isothermal water, separated by *sheets* 0.1-0.3 m thick across which the temperature changes sharply. The sheets are regions of strong shear; the layers, regions of weak shear. (Shear is discussed in Sec. 4.3). The divers have been able to photograph internal waves with wavelengths of 5-10 m and periods of about 5 min which deform the sheets. Occasionally, rows of billows will form on a wave crest. They take about two minutes to roll up, then secondary instabilities develop and the initially regular array of billows breaks down into turbulence (Woods, 1969b). Woods (1968) has shown that the shear which is observed at the crests and troughs of the internal waves is sufficient to cause the instability. Woods (1969a) has also observed that in some cases an internal wave will steepen and overturn as a plunging breaker. These breakers are longer and their outline far more ragged than the billow shown in Fig. 1-7. They scarcely complete the first overturn, let alone roll up. This type of instability occurs at the top of the thermocline during the morning after the

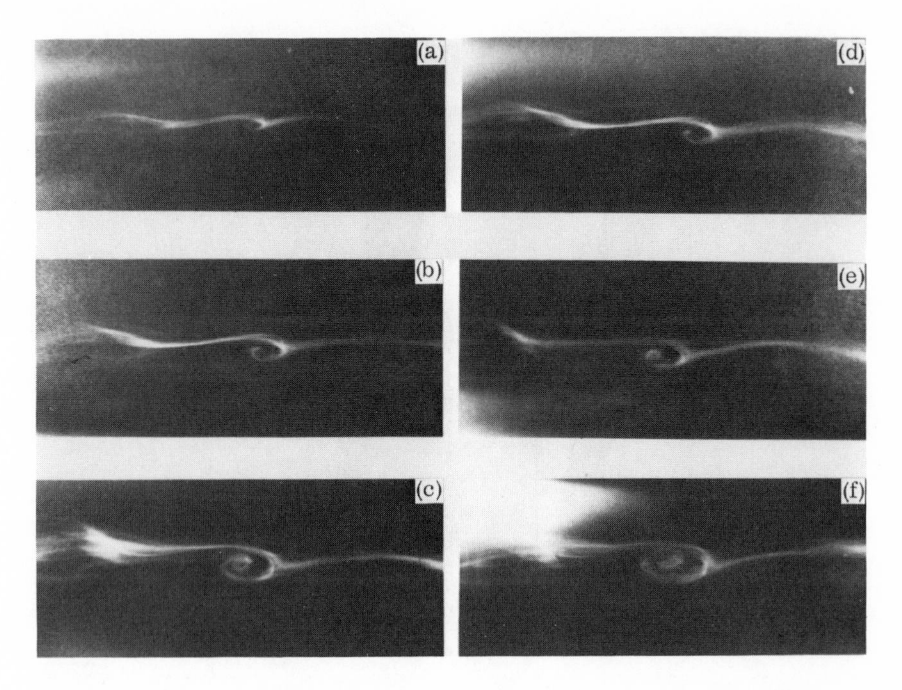

FIG. 1-7. Billow forming on the crest of an internal wave. The wave depicted here has a wavelength of 2.5 m and an amplitude of 0.3 m. (Reprinted from Woods, 1968, by permission of Cambridge University Press.)

temperature gradient across the top sheet has been weakened by the surface heat loss during the night.

By using the procedure described below, Thorpe (1968a, 1971b) has been able to reproduce in the laboratory the roll or billow forms found by Woods off Malta. A long tube is completely filled with equal depths of fresh water and brine, the brine being colored with dye. The tube is left in a horizontal position until the required density profile is obtained as a result of molecular diffusion. Then the tube is suddenly tilted through a small angle (about 8°). The fluid accelerates under gravity and instability due to shear (Thorpe, 1969a; 1971b) occurs after about 3 sec (Fig. 1-8). Thorpe has obtained another shape for breaking waves by filling the tube from the bottom, when its longest side is vertical, with two layers of fluid separated by a sharp interface. The tube is then rapidly tilted into

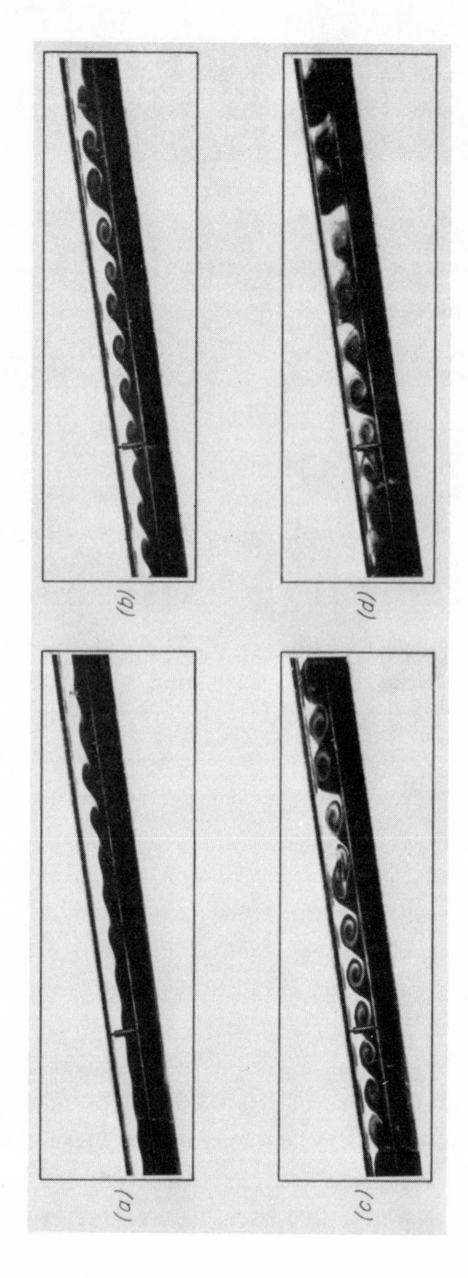

FIG. 1-8. Sequence at approximately half-second intervals showing the growth of rolls, or billows, at the interface between two fluids of equal depth in relative acceleration. (Redrawn from Thorpe, 1968a; used with the author's permission.)

a horizontal position, and breaking waves form on the interface (Fig. 1-9). These cusped waves result from a shear instability at the edge of the interface; such waves may result when the thickness of the density interface is less than that of the velocity interface. Hazel (1972) has given some theoretical examples (Thorpe 1972: personal communication). In a recent paper, Thorpe (1973) has reviewed laboratory experiments on turbulence and breaking internal waves.

Shoaling internal waves in the ocean have been observed by Cairns (1967), who found that as the internal tide propagates into water of depth comparable to its amplitude, its leading edge becomes increasingly steep and its trailing edge flatter. A very recent paper by Emery and Gunnerson (1973) deals with the breaking of internal waves in Santa Monica Bay, California. While the breaking of shoaling internal waves in the ocean has been only indirectly indicated (Wunsch and Dahlen, 1970a; Summers and Emery, 1963), tank experiments give a clue as to what may actually be going on. The recent work by Cacchione (1970) and Hall and Pao (1969, 1971) is particularly helpful.

Cacchione (1970) has considered the shoaling of both high- and low-frequency internal waves in a linearly stratified fluid. In general, he has found that low-frequency waves over a steep slope and high-frequency waves over a shallow slope both develop considerable turbulence. If the slopes are equal, the low-frequency waves tend to break less violently and typically form surges (Fig. 1-10).

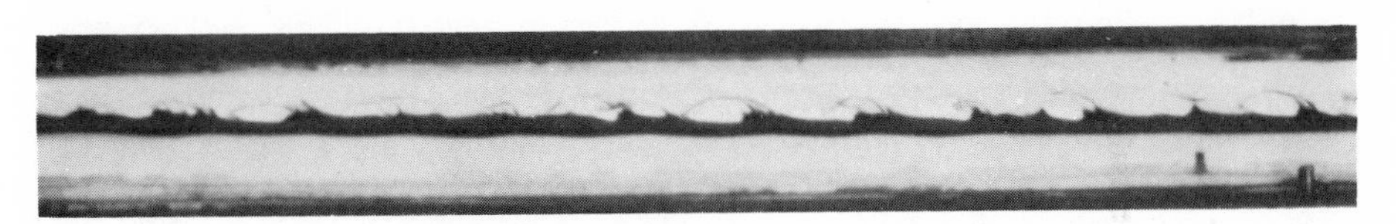

FIG. 1-9. Cusped waves. The upper fluid is moving left and the lower fluid is moving to the right. The wavetrain is moving to the left. The dye belongs to fluid of intermediate density. (Reprinted from Thorpe, 1968a, by permission of Cambridge University Press.)

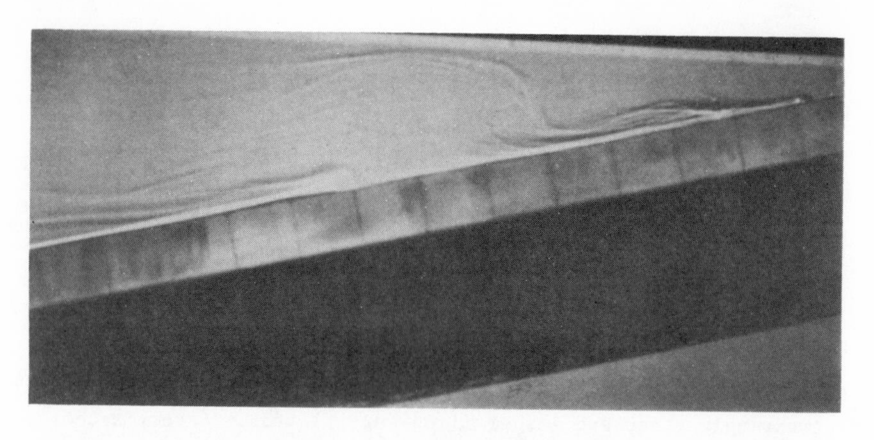

FIG. 1-10. Low-frequency ($\omega/N = 0.375$) internal wave over a slope of 15°. Time elapsed from the start of the wavemaker is 5 min, 15 sec. (From Cacchione, 1970; used with the author's permission.)

Fig. 1-11 shows an internal wave over a steeper slope. The mixing becomes intense after the collapse of the initial vortex [Fig. 1-11 (a)]. Note the reversal of the circulation of the lead vortex as it propagates upslope and collapses [direction of the circulation is indicated by the arrows in Figs. 1-11(a)-(c)]. The bottom effects (Fig. 1-12) have led Cacchione to suggest that internal waves may be important in sediment movement.

Hall and Pao's (1971) work is more concerned with the actual mechanism which causes the waves to break. They have considered long waves shoaling on a gradually sloping beach in a two-fluid system. In all of the cases which they observed, the breakdown of the primary shoaling wavetrain begins with the formation of small ripples on the wave crest (Fig. 1-13), which Hall and Pao have shown is most probably due to shear instability. Then, depending upon the amplitude of the primary wavetrain, there are two ways the wave can continue to break: (a) The ripples themselves can break into smaller ripples which cover the entire crest, so that the interface is highly random (Fig. 1-14); or (b) the wave can form a bore, with the

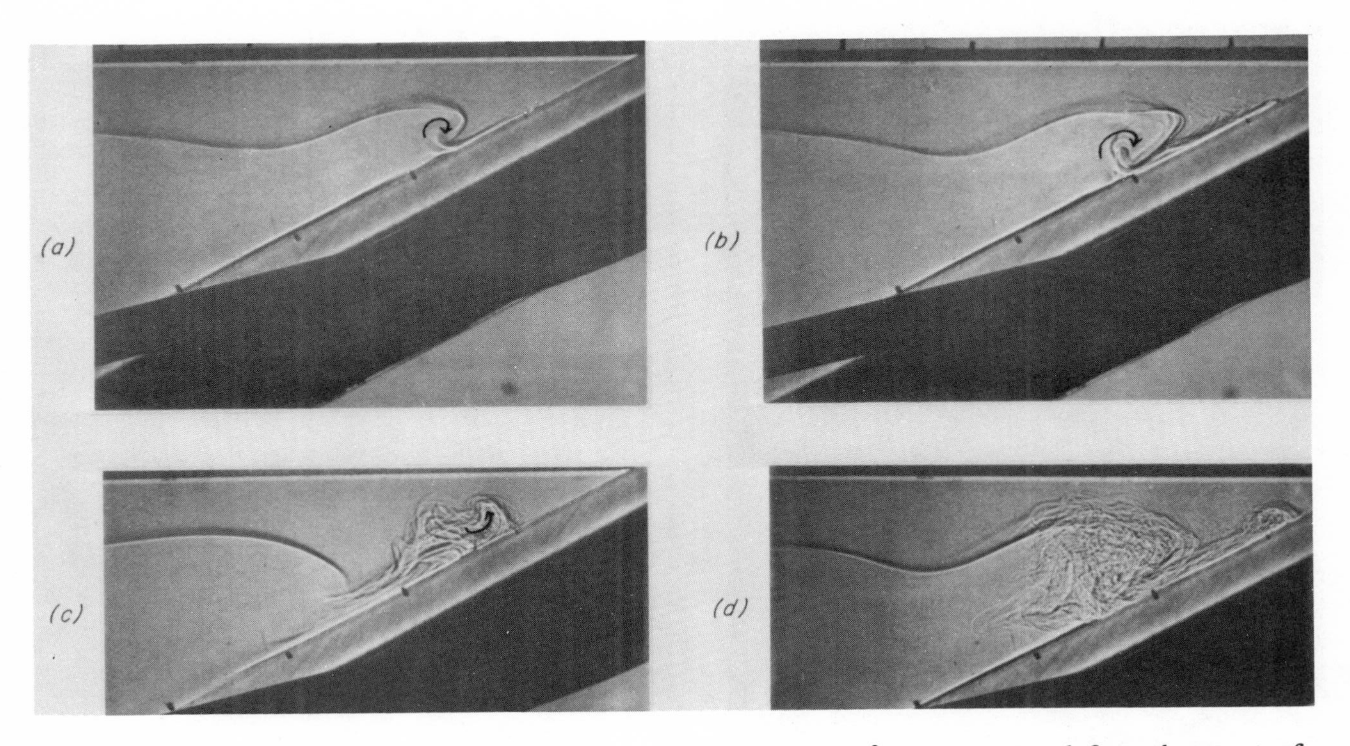

FIG. 1-11. Internal wave with $\omega/N = 0.6$ over a slope of 30°. Time elapsed from the start of the wavemaker: (a) 2 min, 20 sec; (b) 2 min, 30 sec; (c) 2 min, 40 sec; (d) 2 min, 55 sec. (From Cacchione, 1970; used with permission from the author.)

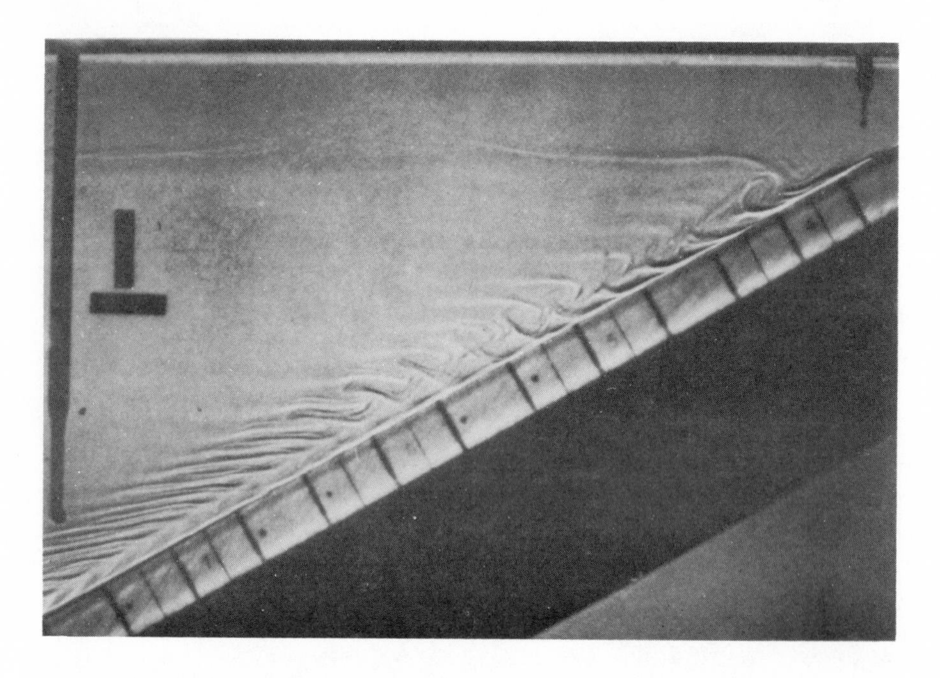

FIG. 1-12. Internal wave with $\omega/N = 0.46$ over a slope of 30°. Time elapsed from start of wavemaker not given. (From Cacchione, 1970; used with the author's permission.)

ripples becoming so large that there is a horizontal detachment of the wave crest due to the shearing motion (Fig. 1-15). The second process is the dominant one for highly steepened waves. Hall and Pao have also considered the breaking of internal waves at the interface between two miscible fluids.

Nagashima (1971) has considered the reflection of internal waves by a sloping beach in a wavetank filled with two immiscible fluids. His results on the types of breaking agree qualitatively with those of Hall and Pao (1971). He has given some reflection coefficients and has discussed the various types of breaking for different beach slopes.

Thorpe (1968b) has found that if the amplitude of a standing internal wave in a two-fluid system becomes too large, the wave will break, not at the crest, but at the nodes where the shear is great-

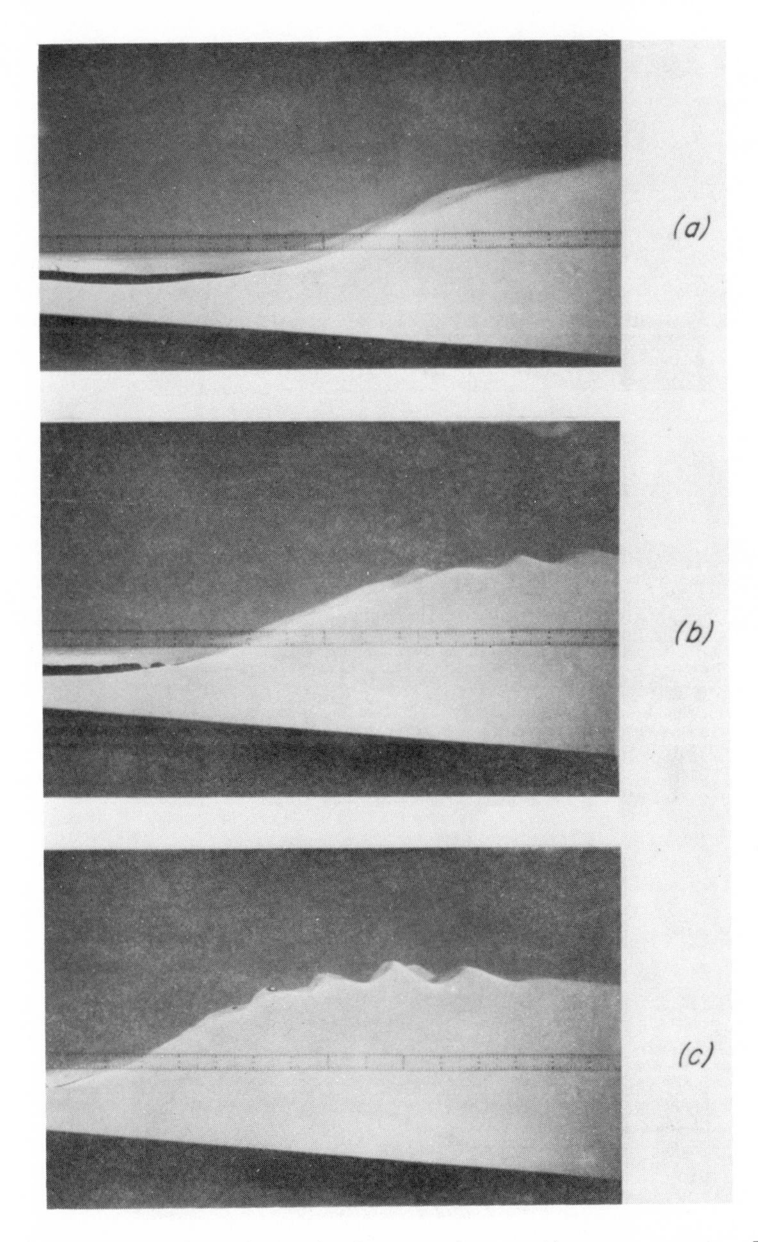

FIG. 1-13. Secondary ripples growing on the wave crest. The wavelength of the large incident wave is 2.58 m. Its amplitude is 0.05 m. The slope is 4°. Times are not given. (From Hall and Pao, 1971; used with Dr. Pao's permission.)

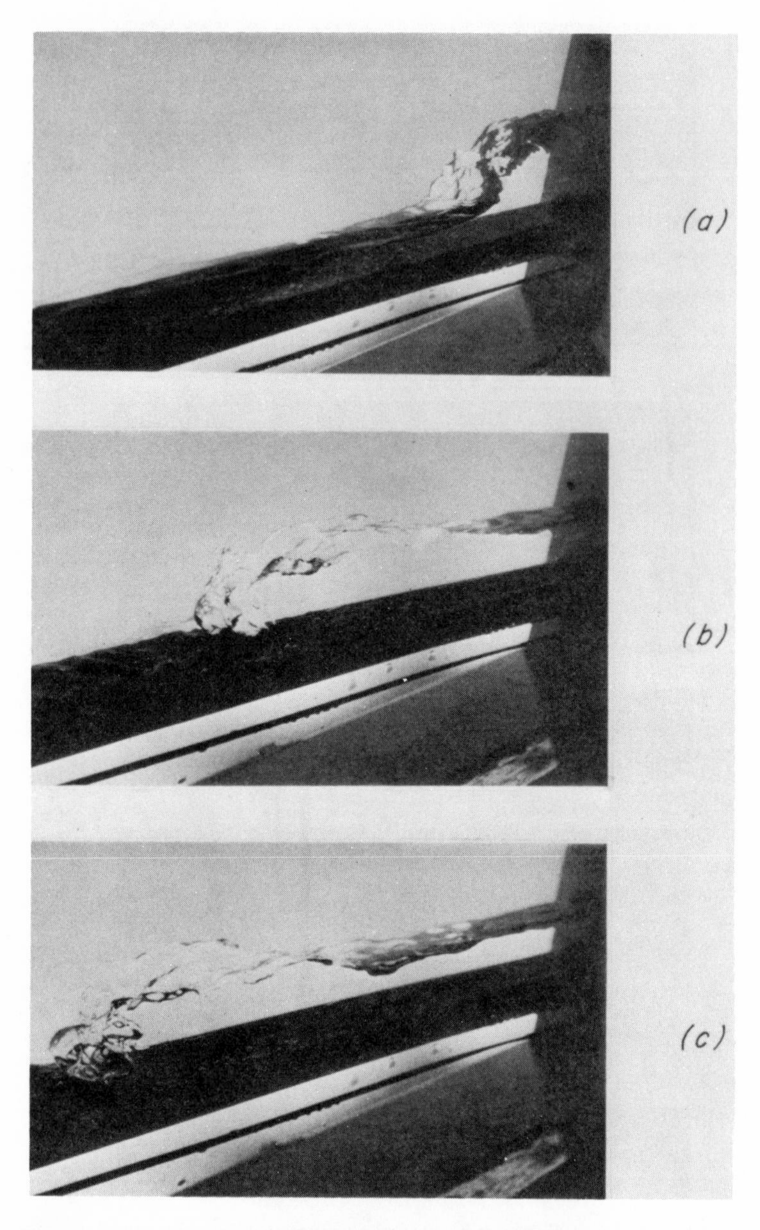

FIG. 1-14. Sequence showing the development of random disturbances of the interface. Wavelength of incident wave is 2.67 m, its amplitude is 0.04 m. The slope is 4°. Times are not given. (From Hall and Pao, 1971; used with Dr. Pao's permission.)

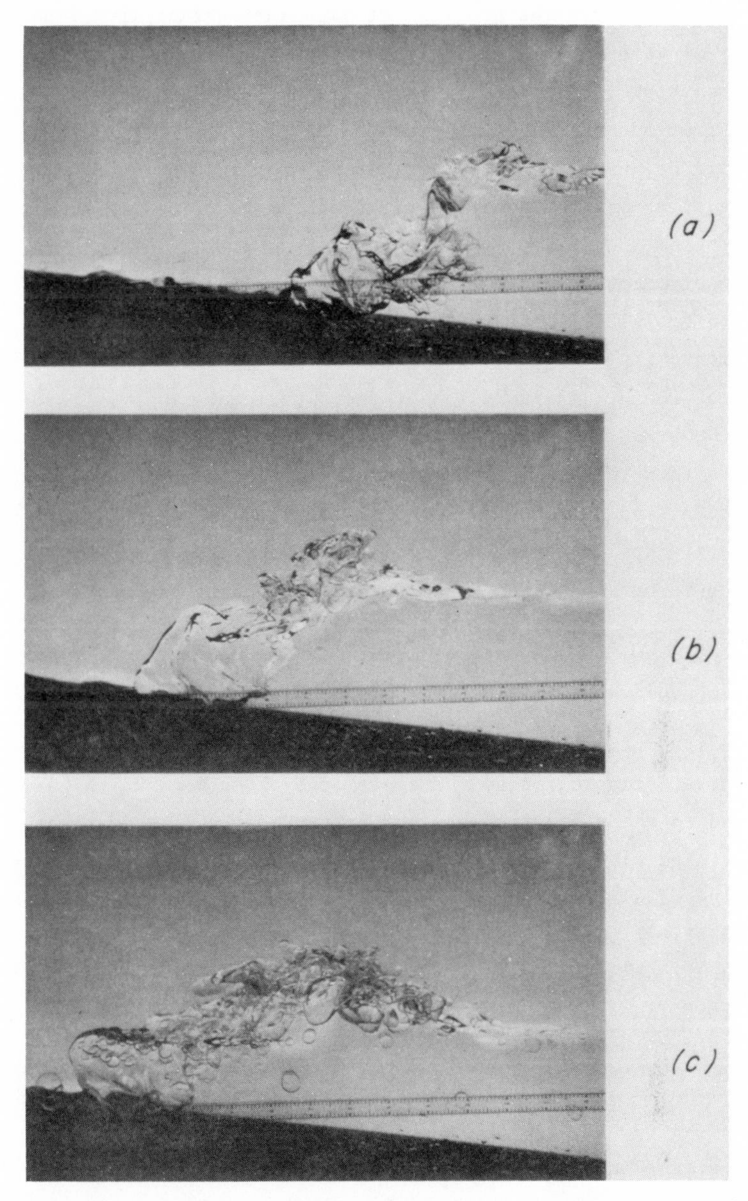

FIG. 1-15. Sequence showing more violent breaking. Incident conditions are the same as in Fig. 1-14. The slope is now 8°. (From Hall and Pao, 1971; used with Dr. Pao's permission.)

est (Fig. 1-16). The progress of resonant instability of a standing internal wave in a linearly stratified fluid is shown in Fig. 1-17. A discussion of McEwan's (1971) wavetank experiments from which these illustrations are taken is given in Sec. 4.2. Phillips, George, and Mied (1968) have found an increase in amplitude of an internal wave propagating against a current. If the current becomes too large, the wave becomes unstable and turbulence is generated, though no tumbling is observed at the wave crests or troughs.

Fig. 1-18 illustrates Deardorff, Willis, and Lilly's (1969) experiment simulating the lifting of an atmospheric inversion above heated ground. A discussion of the experiment is given in Sec. 4.5.

Radar sounder records of breaking internal waves in the atmosphere show the same roll or billow forms observed by Woods and Thorpe. Wind velocity measurements indicate that the breaking is primarily due to shear instability. Figs. 1-19 and 1-20 show some of the records which have been made from the atmosphere.

1.9 EFFECTS AND INDIRECT INDICATORS OF INTERNAL WAVES

While internal waves are usually detected by temperature, salinity, or current fluctuations, there are other phenomena which have been linked to their occurrence. For instance, the fact that the currents associated with internal waves in a two-fluid system may cause marked slowing of a ship passing through such waters (Ekman, 1904) gave one of the first indications that internal waves may exist in the ocean. Another indication of internal waves was given by Zeilon (1934), who reported that the water for the city of Gothenburg, Sweden, which was supplied from the river Göta älv had at one time been "disagreeably contaminated by salt-water entering from the sea in the form of submarine waves or breakers."

Slicks may be associated with oceanic internal waves, as has been well documented by LaFond (for instance, LaFond, 1962). In Georgia Strait, and presumably in other waters where the same conditions prevail, fresh muddy water overlies more saline, clearer water,

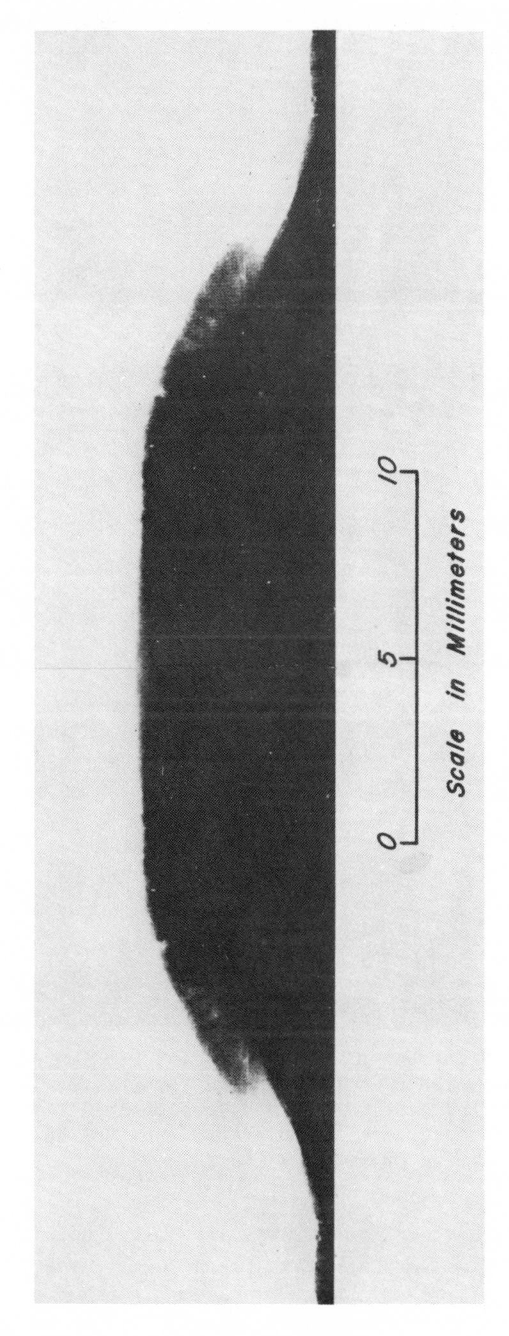

FIG. 1-16. Instability at the interfacial node of a standing internal wave. (Adapted from Thorpe, 1968b.)

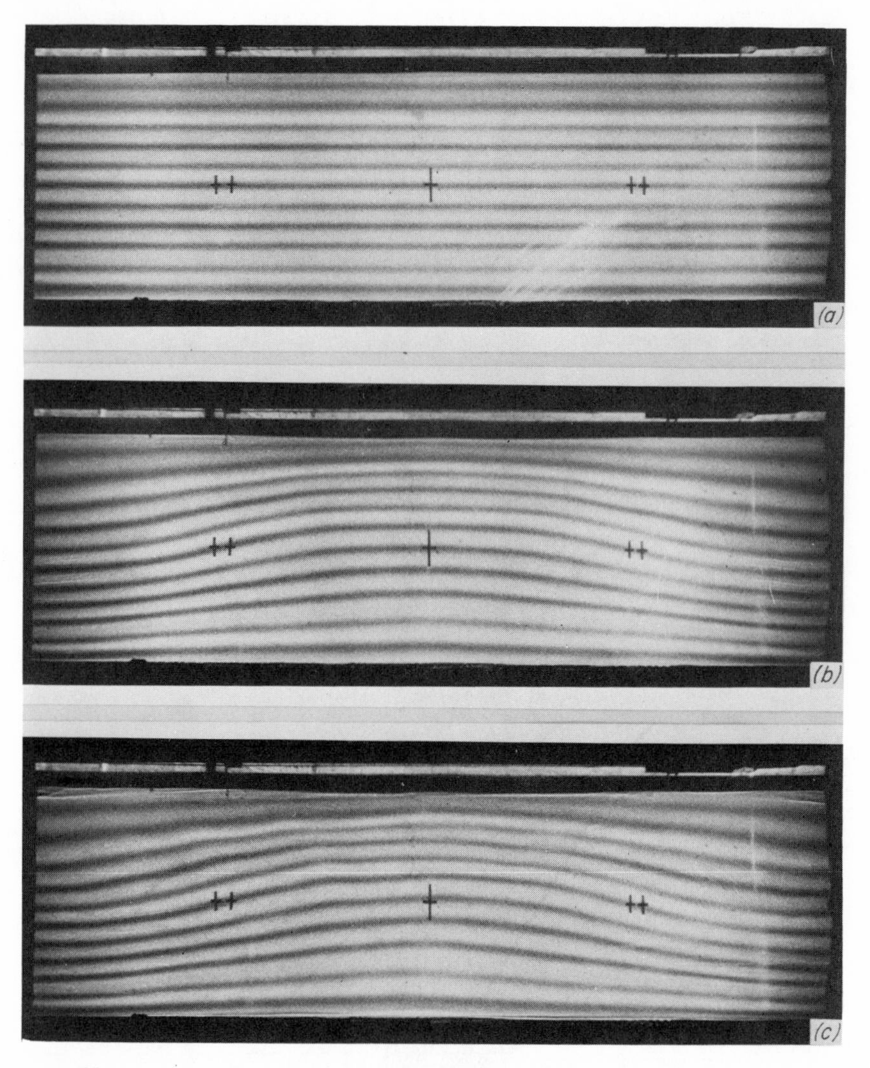

FIG. 1-17. Breakdown of a standing internal wave, $\omega/N = 0.5$. Time is given in cycles of oscillation after the forcing starts: (a) before start; (b) 20 cycles; (c) 45 cycles; (d) 55 cycles; (e) 60 cycles; (f) 65 cycles. (Reprinted from McEwan, 1971, by permission of Cambridge University Press.)

so that the troughs of internal waves are filled up with muddy water and appear as darker bands on the surface (Shand, 1953; LeBlond 1972: personal communication).

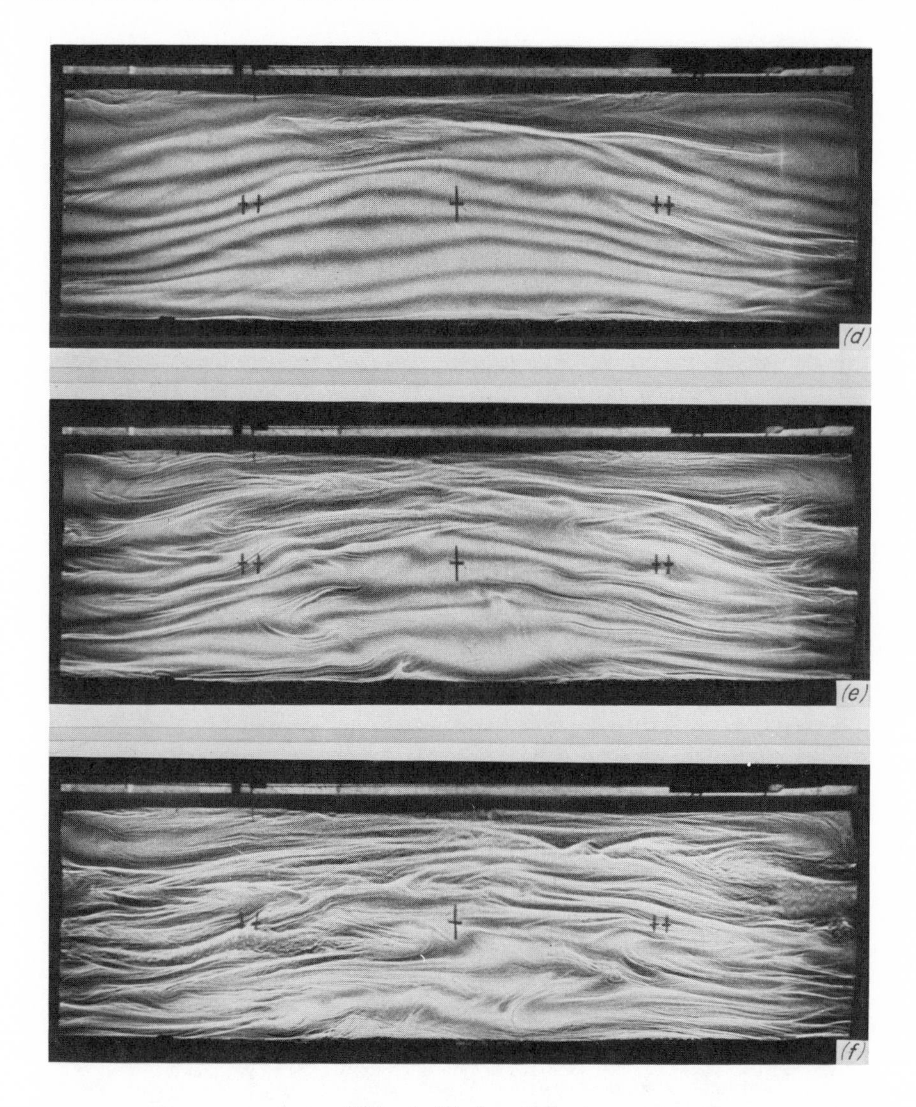

FIG. 1-17 (cont.).

If the thermocline lies sufficiently close to the surface, or if the amplitudes of internal waves are unusually large, they may even cause breakers to appear at the surface. LaFond (1966) has told of seeing bands of small, breaking whitecaps in the Andaman Sea which "emitted a low roar as they passed a drifting ship on a calm sea."

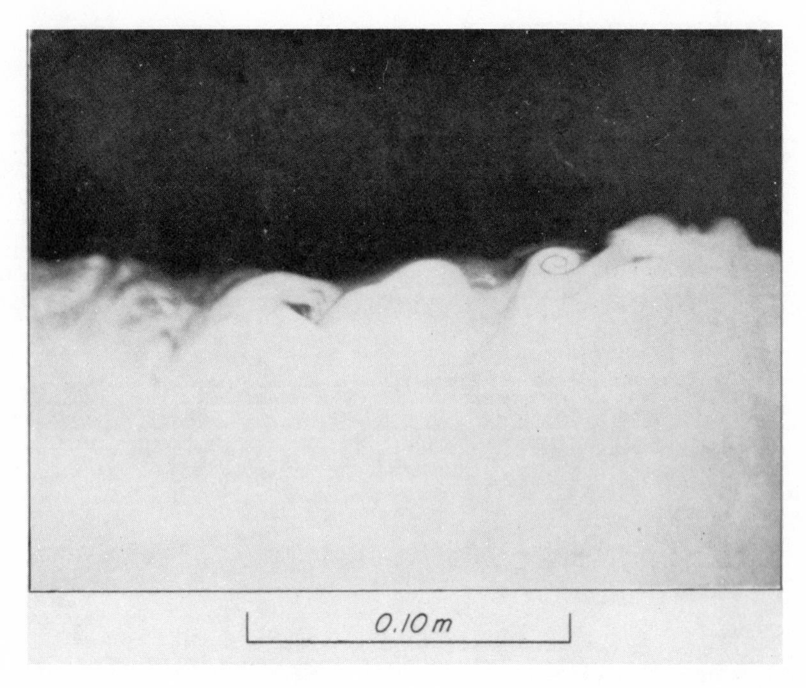

FIG. 1-18. Close-up of the convective region and interface in a tank of width 0.5 m and height 0.35 m. Mean interface height is about 0.17 m. (Reprinted from Deardorff, Willis, and Lilly, 1969, by permission of Cambridge University Press.)

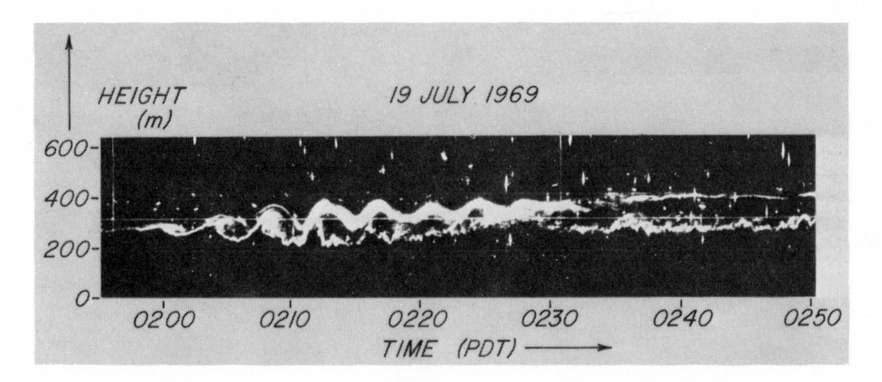

FIG. 1-19. Radar time-height record of atmospheric internal waves taken 19 July 1969 near San Diego. (Reprinted from Atlas *et al.*, 1970, by permission of the American Meteorological Society.)

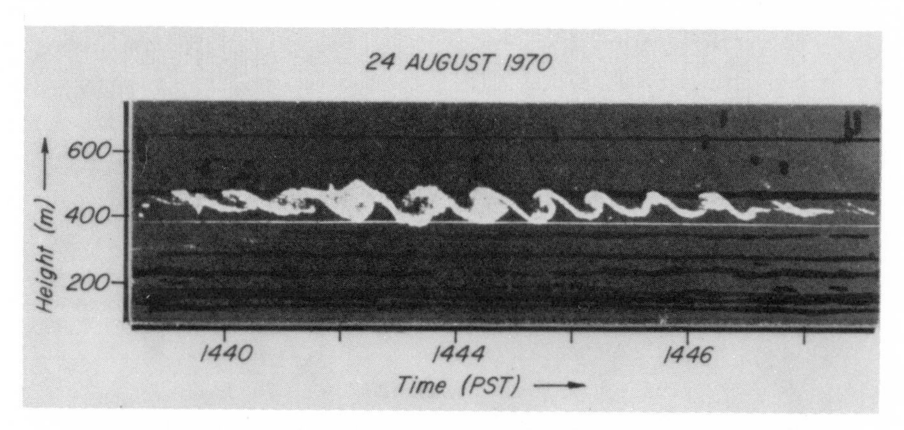

FIG. 1-20. Radar time-height record of breaking atmospheric internal waves taken 24 August 1970, probably near San Diego. Note that the time scale is different from that in Fig. 1-19. (Reprinted from Gossard *et al*, 1971, by permission of the American Meteorological Society.)

Perry and Schimke (1965) have also seen bands of breakers in that region. Larsen (1969a) gives an even more dramatic example. West of Cape Blanco, Costa Rica, at about 9°N, 88°W, is an area shown on the navigation charts as the Guardian Bank. According to the charts, this bank rises steeply from a depth of 3600 to 20 m below the surface of the water. The horizontal distance of this rise is less than 10^5 m, and breakers have been sighted in the region. Yet according to research by Scripps Oceanographic Institution and the University of Washington, the bank does not even exist. The Guardian Bank, or Costa Rica Dome as it is called in oceanographic literature, is a region of upwelling, and the thermocline lies only 10 m below the surface. After a storm, Larsen measured internal waves in the thermocline with heights as great as 5 m. Surface swell propagating over the dome and encountering the adverse currents set up by internal waves could cause the bands of breakers which have been seen there. This would agree with Gargett and Hughes (1972) who have shown that surface waves running over internal waves may alter their direction and increase their amplitudes (Sec. 4.10). Hughes (1972: personal communication) has some striking pictures of this phenomenon.

An unusual internal wave effect has been discovered in a wave-tank experiment by Joyce (1972): A surface wave and a growing internal wave can resonantly generate another surface wave whose frequency will, in general, be lower than that of the first surface wave (in agreement with Hasselmann, 1967).

Sediment formations, such as sand waves, may be caused by internal waves (Emery, 1956; Revelle, 1939; Cartwright, 1959), and breaking internal waves may cause sediment movement (Southard and Cacchione, 1972). Wunsch and Dahlen (1970a) have found exceptionally large microstructure gradients over the slope of Bermuda, consistent with the idea that internal waves are breaking there. Meincke (1971b) has suggested that the anticyclonic vortex above the Great Meteor Seamount may be caused by vertical mixing due to shear induced by internal tides. Internal waves may also affect sound-scattering layers in the ocean (Chindonova and Shulepov, 1965).

1.10 ANALYSIS OF INTERNAL WAVE DATA

Because most internal wave data consists of a time series of temperature and/or current measurements, their analysis is, in general, not unique to the study of internal waves. Moreover, the analysis is usually standard and supplies little insight into understanding the physical phenomenon itself. Thus, most papers on internal waves seldom give it more than passing reference. Some exceptions to this are Fjeldstad's (1933) classic "Interne Wellen"; Haurwitz' (1954) analysis of four time-temperature series discussing the probability that the observed variations are caused by internal tides; Haurwitz, Stommel, and Munk's (1959) analysis of a lengthy time series of temperature measurements near Bermuda; Cox's (1962) discussion of coherence of internal waves; Sabinin and Shulepov's (1965) and Sabinin's (1966) papers on short-period internal waves in which various filters and some details of spectral analysis are discussed; the second part of Krauss' (1966b) book; Fofonoff's (1969) discussion of spectral analysis of vector series; the treatment of Doppler-shifted data by

Sabinin (1969), Etienne (1970), and Laykhtman, Leonov, and Miropol'skii (1971); Schott's (1971a) and Schott and Willebrand's (1973) work on directional spectra; and Phillips' (1971) discussion of the effects of microstructure on the internal wave frequency spectrum.

The treatment of time-series data in general, whether it be for biology, geology, astronomy, or oceanography, is a separate, large, and sometimes abstruse field of study. There are many good books on the subject. Much of the theory has been developed to analyze electrical signals, so that many of the terms used in time-series analysis are borrowed from electrical engineering: power spectra, bandwidths, bandpass filters, lag filters, and white noise. Thus, the author has found Lahti (1968) and Lee (1960) to be very helpful introductions. For harmonic analysis, Chapman and Bartels (1940, pp. 546-605), Chap. 16: "Periodicities and Harmonic Analysis in Geophysics," is still quite useful. Jenkins and Watts (1968) have given an excellent account of spectral analysis of time series.

By far the most comprehensive treatment of analysis of internal wave data *per se* (including Doppler-shifted data) is given in the second half of Krauss' book, *Interne Wellen* (1966b, pp. 144-215). Although the fact that the book is in German may deter some readers, Krauss' mathematical developments are exceptionally clear; and English summaries, located at the end of the book, are provided for each section.

CHAPTER 2

GENERATION OF INTERNAL WAVES

2.1 INTRODUCTION

We have some clues about the sources and mechanisms of internal wave generation in the ocean. For instance, internal waves with tidal periods are probably generated by the surface tide at slopes and sills, and it seems likely that internal waves with periods near the inertial period are sometimes caused by wind. Nonetheless, the disquieting fact remains that unequivocal demonstration of a specific generating force for internal gravity waves in the ocean has not yet been given. The problem is doubly complicated by the fact that the information on dissipation rates and wavenumber-frequency relationships is inadequate. Stern (1972: personal communication) has commented that if someone can figure out a way to get energy from the big geostrophic eddies to internal waves, he will be believed, because the eddies contain more than enough energy. An analytic expression for the generating force for short-period internal waves, whatever its physical source, has been given by Hollan (1972).

Krauss (1966b, Sec. 16, "Origins of internal waves") has given a mathematical formulation of the problem. Possible physical sources of internal waves are discussed in the following sections. Table 2-1 lists generating mechanisms not discussed in the text.

TABLE 2-1

Other Generating Mechanisms

Mechanism	System	Reference
Buoyancy flux at sea surface	Ocean	Magaard (1973)
Currents	Ocean	Fjeldstad (1958) Hollan (1972)
Earthquakes	Two-fluid system	Fedosenko and Cherkesov (1968)
Explosions (nuclear or volcanic)	Atmosphere	Kahalas and Murphy (1971) Scorer (1950) Tolstoy and Lau (1971)
Fronts and jet streams	Atmosphere	Hines (1968)
Heat	Atmosphere	
Lifting of an inversion		Blumen and Hendl (1969) Deardorff *et al.* (1969) Testud (1970)
Moving heat source		Kelly and Vreeman (1970)
Rising thermals		Townsend (1964, 1966)
Solar eclipse		Chimonas (1970)
Shear instability	Atmosphere	Hooke (1973) Kaylor and Faller (1972)
	Ocean	Gade and Ericksen (1969) Mork and Gade (1967)
Spherical source which breathes fluid in and out	N constant	Hendershott (1969)
Surface turbulent boundary layer	Atmosphere	Townsend (1965)
	Ocean	Bretherton (1971)

2.2 RESONANT WAVE-WAVE INTERACTIONS

Resonance theory was first applied to surface gravity waves by Phillips (1960) and led to investigations of wave-wave interactions among surface and internal waves. This section opens with a brief background of wave-wave interactions, followed by a discussion of interactions between surface waves and internal waves. Finally, the generation of internal waves by surface waves and by other internal waves is considered. Resonant instability is discussed in Sec. 4.2.

In general, any wave may be characterized by its frequency and its wavenumber; that is, $(\omega,\vec{k})$ represents a certain wave with frequency ω and vector wavenumber $\vec{k}$. For the sake of simplicity, suppose that we have a linear system (e.g., two superposed fluids), describable by equations with constant coefficients having solutions expressible as the sum of undamped waves of the form $\exp[i(\vec{k}\cdot\vec{r} - \omega t)]$. Only certain ω-$\vec{k}$ relationships can hold for a linearized system; i.e., ω and $\vec{k}$ must satisfy a dispersion relation

$$G(\omega,\vec{k}) = 0 \tag{1}$$

Suppose we generate two waves $(\omega_1, \vec{k}_1)$ and $(\omega_2,\vec{k}_2)$, each of which satisfies Eq. (1), in such a manner that they act on the system (e.g., two waves on the upper surface of a two-fluid system). The system will support four waves: the two *primary waves* $(\omega_1,\vec{k}_1)$ and $(\omega_2,\vec{k}_2)$, and two *combination waves* $(\omega_1 + \omega_2, \vec{k}_1 + \vec{k}_2)$ and $(\omega_1 - \omega_2, \vec{k}_1 - \vec{k}_2)$. We have $G(\omega_1,\vec{k}_1) = G(\omega_2,\vec{k}_2) = 0$, but in general, $G(\omega_1 \pm \omega_2, \vec{k}_1 \pm \vec{k}_2) \neq 0$.

When the combination waves interact, they generate other waves called *secondary*, or *forced, waves* with wavenumbers $\vec{k}_1 + \vec{k}_2$ or $\vec{k}_1 - \vec{k}_2$. Because of the dispersion relation, i.e., because of Eq. (1), the secondary waves must have a frequency ω_j which satisfies $G(\omega_j, \vec{k}_1 + \vec{k}_2) = 0$ or $G(\omega_j, \vec{k}_1 - \vec{k}_2) = 0$, so that they propagate with phase speed

$$c_p = \frac{\omega_j}{|\vec{k}_1 \pm \vec{k}_2|} \tag{2}$$

In general, these secondary waves will be damped out almost as quickly as they are generated. However, if one of the secondary waves, either $(\omega_j, \vec{k}_1 + \vec{k}_2)$ or $(\omega_j, \vec{k}_1 - \vec{k}_2)$, satisfies $G(\omega_3,\vec{k}_3) = 0$, where $(\omega_3,\vec{k}_3) = (\omega_1 \pm \omega_2, \vec{k}_1 \pm \vec{k}_2)$, then this secondary wave is coherent with the combination wave and we have *primary resonant interaction*. To reiterate, the condition for resonant interaction between the primary waves $(\omega_1,\vec{k}_1)$ and $(\omega_2,\vec{k}_2)$ is that

$$G(\omega_1,\vec{k}_1) = G(\omega_2,\vec{k}_2) = G(\omega_3,\vec{k}_3) = 0 \tag{3a}$$

where

$$\omega_3 = \omega_1 \pm \omega_2 \tag{3b}$$

$$\vec{k}_3 = \vec{k}_1 \pm \vec{k}_2 \tag{3c}$$

When Eqs. (3) hold, the three waves, $(\omega_1,\vec{k}_1)$, $(\omega_2,\vec{k}_2)$, and $(\omega_3,\vec{k}_3)$ are said to form a *resonant triad*. A continuous transfer of energy is possible among the three waves, and the momentum and wave energy of the resonant triad are constant. It is also possible to have resonant interaction between four waves, $(\omega_1,\vec{k}_1)$, $(\omega_2,\vec{k}_2)$, $(\omega_1 - \omega_2, \vec{k}_1 - \vec{k}_2)$, and $(\omega_1 + \omega_2, \vec{k}_1 + \vec{k}_2)$, if $G(\omega_1,\vec{k}_1) = G(\omega_2,\vec{k}_2) = G(\omega_1 - \omega_2, \vec{k}_1 - \vec{k}_2) = G(\omega_1 + \omega_2, \vec{k}_1 + \vec{k}_2) = 0$. However, in physical systems the equation $G(\omega_1 + \omega_2, \vec{k}_1 + \vec{k}_2) = 0$ does not often hold, and only the case of three interacting waves is usually considered.

The differences between interactions in a homogeneous and in a stratified fluid have been discussed by Thorpe (1966). Simmons (1969) has considered weak, resonant wave interactions taking frequency and wavenumber to be constant, and slowly modulating the wave amplitude and phase angle, a method which simplifies the interaction equations.

A general energy equation for resonantly interacting random wave components has been given by Hasselmann (1966), who has applied concepts from plasma physics and quantum mechanics to geophysical scattering problems. One of the main contributions of this paper is to point out that the theory of random wave-wave interactions of quantum mechanics is applicable to oceans. Hasselmann has worked out the coupling coefficients for several different wave fields, including the coefficient for the interactions of surface and internal waves. After presenting scattering theory in a Hamiltonian form, the energy transfer due to weak nonlinear interactions in random wave fields is reinterpreted in terms of a hypothetical ensemble of interacting particles. The particle picture is used only as an abstract description of the effects of wave interactions; it permits wave scattering to be compactly represented by Feynman, or transfer, diagrams. (For Hasselmann's paper to be thoroughly understood, the reader needs a background in intermediate quantum mechanics, this author not being so qualified.) Feynman diagrams have been used by Kenyon (1968) to represent surface-surface, surface-internal, and internal-internal wave interactions in the ocean. He has found that the time scale for the interaction of surface waves and internal waves is much larger than the time scale either for mutual internal wave interactions or for mutual surface wave interactions (except in the unimportant case of very weak interactions).

It may be shown that in a two-fluid system, two surfaces waves can interact resonantly to produce an internal wave at the interface (Ball, 1964; Nesterov, 1970; Joyce, 1972). The internal wave thus generated will have an amplitude which increases exponentially, its growth exponent being proportional to the initial amplitude of the surface wave of highest frequency (Nesterov, 1970). Ball (1964) has stated that, for resonant interaction to occur, the two surface waves must be moving in opposite directions, but Thorpe (1966) has pointed out that theoretically, one could have interactions between three waves all traveling in the same direction. This, however, would require that the group velocity of one of the surface waves be less than the limit of the group velocity of the internal wave as its wavenumber tends to zero. Hence, it seems unlikely that this case is of much practical importance.

It is also possible for two surface waves to generate an internal wave in a fluid with a continuous density profile (Thorpe, 1966;

Kenyon, 1968; Krauss, 1967; Phillips, 1966; Joyce, 1972). In a wave-tank, the generated internal wave may have an amplitude equal to that of the surface waves after an interaction time of only about half a minute (Thorpe, 1966). In general, it is necessary that the two surface waves be traveling in the same direction, or in nearly the same direction (Kenyon, 1968; Phillips, 1966, p. 173). If the thermocline depth -d is significantly greater than the wavelengths of the surface waves (as is usually the case in the ocean), then the internal wave will have a wavelength of the order d or greater, and will propagate almost at right angles to the mean direction of the surface waves (Phillips, 1966, p. 173).

The question of whether internal waves actually are resonantly generated in the ocean is so far unanswered. The use of an exponential density model (N constant) seems to indicate that the resonant effect is small in the deep ocean (Kenyon, 1968). However, a density model which includes a thermocline should permit a much larger transfer of energy from surface waves to internal waves (Joyce, 1972), and in shallow water the energy transfer may be greater yet (Kenyon, 1968).

In fact, preliminary investigations seem to indicate that internal waves with 2-20 min periods which occur off the coast of Southern California are indeed resonantly generated by surface waves (Krauss, 1967). Since both surface swell and internal waves usually travel toward shore off the coast of Southern California, the idea of internal-wave generation by surface waves is in seeming contradiction to Kenyon (1968) and Phillips (1966, p. 173). Krauss (1967), however, has considered a case where the swell may be described by two surface waves, $(\omega_1,\vec{k}_1)$ and $(\omega_2,\vec{k}_2)$, with amplitudes a_1 and a_2, respectively, (i.e., the swell is amplitude-modulated) traveling in directions which make an angle with each other between 90° and 180°. If $a_1 \gg a_2$, this still gives a progressive wave which will travel nearly in the direction of the larger wave. For this case, Krauss has found that internal waves with amplitudes of 1-3 m can be created within 15 min, if the stratification of the water is such that

$$\nu = \frac{|\vec{k}_1 - \vec{k}_2|^2}{(\omega_1 - \omega_2)^2}$$

is an eigenvalue of the internal-wave equation (when the earth's rotation is neglected), in agreement with Eq. (2), because

$$\nu = \frac{1}{c_p} \tag{4}$$

[See Sec. 3.4, particularly Eq. (3.4.11).] These internal waves will have the same wavelength and period and travel in the same direction as the amplitude modulation. Preliminary results from simultaneous measurements of surface swell and internal waves at the NEL Tower off the coast of Southern California made in October 1966, have supported the theory. [Krauss' (1967) report contains clear derivations, using perturbation equations, of the equations of motion through second order. It also includes an expression for the second-order vertical velocity equation in terms of two driving forces, f_1 and f_2, responsible for second-order internal waves; f_1 is the body force in the fluid due to nonlinear terms in the equations of motion and the incompressibility condition; f_2 is a surface force due to the deviations of the sea surface from its mean level.]

Internal waves may also be resonantly generated by other internal waves, although this has been experimentally confirmed only in wavetanks. The strongest mutual internal wave interactions in the ocean probably take place in the relatively shallow water near the coast (Kenyon, 1968).

If we assume that the Brunt-Väisälä frequency N is constant, that the surface is rigid, and that the water depth is -D, the resonance conditions are

$$\omega_3 = \omega_1 \pm \omega_2 \tag{5a}$$

$$\vec{k}_{h_3} = \vec{k}_{h_1} \pm \vec{k}_{h_2} \tag{5b}$$

$$n_3 = n_1 \mp n_2 \tag{5c}$$

(Kenyon, 1968), where $\vec{k}$ is a horizontal wavenumber vector given by

$$\vec{k}_h = \hat{i}k_x + \hat{j}k_y$$

and n is a mode number, related to the vertical wavenumber k_z by

$$k_z = \frac{n\pi}{D} \tag{6}$$

Because of Eq. (5c), interaction cannot occur among three internal waves belonging to the same mode (Thorpe, 1966). For short waves when N is constant, the dispersion relation is given by

$$G(\omega,\vec{k}) = \left(\frac{\omega}{N}\right)^2 - \frac{k_h^{\,2}}{k_h^{\,2} + k_z^{\,2}} = 0 \tag{7}$$

[See Sec. 3.4, particularly Eqs. (3.4.19) and (3.4.23).]

Martin, Simmons, and Wunsch (1969) have shown that two internal waves, $(\omega_1,\vec{k}_1)$ and $(\omega_2,\vec{k}_2)$, can generate a third internal wave at the difference frequency $\omega_1 - \omega_2$, where the three waves satisfy Eqs. (5). Even one internal wave, if its amplitude is sufficiently large, can resonantly generate other internal waves in a wavetank (Martin, Simmons, and Wunsch, 1972). If a resonant triad is generated in this way, the frequencies of the two secondary internal waves are always less than the frequency of the original wave (Hachmeister and Martin, 1973). Martin, Simmons, and Wunsch (1972) have been able to generate as many as five resonant triads from a single primary internal wave, each triad having the original internal wave as one of its members, leading them to conclude that "resonant interactions provide a powerful mechanism for transferring energy from a single driving frequency to other parts of the (frequency) spectrum."

2.3 WIND AND AIR PRESSURE FLUCTUATIONS

The idea that the wind might cause internal waves was first proposed by Watson in 1904 to explain internal seiches in Loch Ness. He wrote, "Since winds produce a tilt in the isotherms, and the stronger the wind the greater the depth to which its effect is felt, it seems probable that these swingings are started by strong winds which are able to displace the isotherms in the critical region." The idea was confirmed experimentally a few years later by Standström (1908), who found he could produce internal waves in a tank by blowing gently and periodically on the surface. Further wavetank experiments were done by Zeilon (1934) and Mortimer (1951). Fig. 2-1 from Mortimer's experiments shows how the wind may act to produce an internal seiche in a lake.

As experimental evidence mounted for some sort of connection between storm systems and internal waves in the Baltic and the North Atlantic, efforts were renewed to obtain a workable model for wind generation of internal waves. Tomczak (1966, 1967a, 1967b) has shown that both divergence and curl of the wind field can cause internal

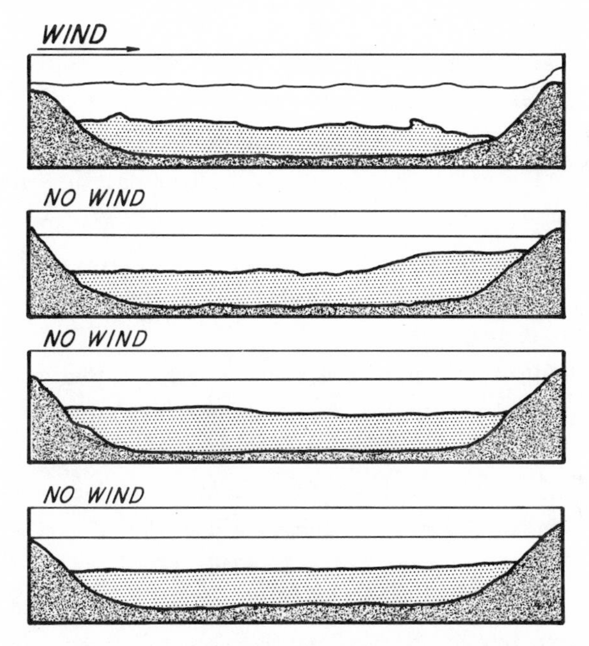

FIG. 2-1. Successive stages of an internal seiche caused by wind on the surface of a two-layered "lake." (Redrawn from Mortimer, 1951.)

waves in an exponentially stratified ocean. Krauss (1972) and Mork (1968b) have also considered the effect of wind stress on a continuously stratified ocean. The following discussion of response functions is due to Krauss (1972).

A linear system acted on by a force $F(t)$ may be described by a differential equation

$$Dy(t) = F(t) \tag{8}$$

where $y(t)$ is a variable, such as displacement or velocity of the system, and D is a linear differential operator. A solution of Eq. (8) in terms of a Green's function $G(t)$ is

$$y(t) = \int G(t-\xi)F(\xi)d\xi$$

In the frequency domain, this solution reduces to the algebraic expression

$$\tilde{y}(\omega) = R(\omega)\tilde{F}(\omega)$$

where $\tilde{y}(\omega)$ and $\tilde{F}(\omega)$ are the Fourier transforms of $y(t)$ and $F(t)$. (For instance, see Lee, 1960, Chap. 13). $R(\omega)$ is called the *response* (or *transfer*) *function* and is the Fourier transform of $G(t)$. From a physical standpoint, $R(\omega)$ is the Fourier spectrum of $y(t)$ for an input force $F(t) = \delta(t)$, where $\delta(t)$ is a unit-impulse function. The response functions due to wind stress and air pressure must be multiplied by the wavenumber-frequency spectra of the wind stress and air pressure, respectively, in order to obtain an actual displacement or current response.

Krauss (1972) has obtained expressions for the response functions due to wind stress and air pressure, and has displayed them in the wavenumber-frequency plane for various depths. One such graph is shown in Fig. 2-2. Krauss has found that wind-induced vertical displacements within the water column are due almost entirely to internal waves. Inertial waves are a dominant feature for wind response in shallow as well as deep water, and internal waves and long surface waves are more important than quasipermanent currents and displacements. Krauss has concluded that wind-induced velocity fluctuations

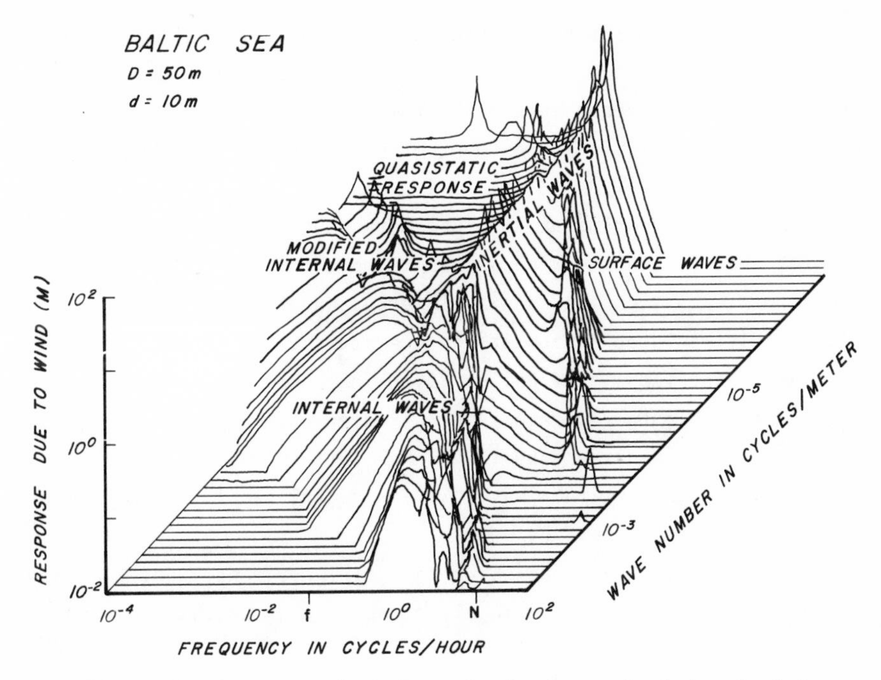

FIG. 2-2. Response functions for horizontal wind velocities. (Redrawn from Krauss, 1972, and used with the author's permission.)

have amplitudes at least as large as those of wind-produced mean currents.

Magaard (1971) has given a model for a continuously but arbitrarily stratified ocean which compares the models of Tomczak, Mork, and Krauss. Magaard's model differs from Tomczak's, which uses a rigid surface; from Mork's, which requires that the horizontal velocity be proportional to shear stress; and from Krauss', for which the exchange coefficients are constant. To apply Magaard's model to the real ocean, one needs the air pressure and wind spectra for the corresponding space and time domains. Work is now in progress to obtain this information for the Baltic Sea (Kielmann 1972: personal communication).

Generation of inertial waves by wind has been discussed theoretically by Pollard (1970) and by Pollard and Millard (1970). Crepon (1971, 1972) has used a two-layered ocean and discontinuous variations in wind intensity (caused by a coast, for example) to explain the generation of near-inertial oscillations with amplitudes of a few meters. Heaps and Ramsbottom (1966) and Csanady (1968) have given a theoretical treatment of wind generation of internal waves in lakes.

The models of Tomczak (1967b), Mork (1968b), and Krauss (1972) predict that tangential wind stress is much more effective than air pressure fluctuations in causing internal waves, although Mork points out that it would be difficult to separate the two in oceanic observations. Käse (1971) and Faller (1966) have shown that it is theoretically possible to generate internal waves with air pressure fluctuations alone, and Keunecke (1970) has in fact experimentally produced standing internal waves in a wavetank simply by varying the air pressure at the surface.

Polyanskaya (1969) has shown that a plane front squall, accompanied by a local variation in air pressure and increase in wind velocity, and propagating with velocity $\vec{U} = \hat{i}U_x$, may generate internal waves whose wavelengths are such that their phase velocities are equal to $\vec{U}$. The waves may have amplitudes of several meters.

Since wind-generated surface waves may in turn generate internal waves by resonant interaction (Sec. 2.2), storms may be an indirect generator of internal waves (Hasselmann, 1970).

2.4 TOPOGRAPHY

Seamounts and sills

In analogy with lee waves in the atmosphere, internal waves may be produced when a current in the ocean flows over an irregularity on the ocean floor, such as a sill or a seamount. Cox and Sandstrom (1962) have started with the idea that all internal waves which are produced by a bottom roughness possess the same period as the primary wave (which may be the surface tide or another internal wave). They have shown that a plane surface wave propagating over a rough patch on the ocean floor produces a system of internal waves which radiate outward from this region. The amplitudes of these waves are proportional to the amplitudes of the surface wave. Krauss (1966b, pp. 112-115) has given a clear summary of the mathematical development of Cox and Sandstrom, and their results have been extended by Hendershott (1968). Using data from the open Pacific, Bell (1973a) has shown that a bottom current flowing at 4 cm/sec over small scale topography can result in a transfer of energy from the current into the wave field whose magnitude is the same as the energy transferred from the wind to the ocean currents. "The implications of this study," Bretherton (1972: personal communication) has commented, "are hard to escape."

Zeilon (1912) has demonstrated both experimentally and theoretically the effects of a sill on the generation of the internal tide. Generation of long internal waves by tidal flow over a bottom irregularity has been considered by Mork (1968a) for the case when the phase speed c_p of the waves is less than or equal to the maximum total current U_{max} (Zeilon, 1912, required that $U_{max} \ll c_p$). The flow over a square sill for $U_{max} < c_p$ is shown in Fig. 2-3a, and over a smooth bump for $c_p < U_{max}$ in Fig. 2-3b. When $1/2 < c_p/U_{max} < 1$, the total vertical displacement of the water particles in a tidal current is about twice the height of the bottom irregularity. The wave slope becomes very steep as c_p approaches U_{max}. If c_p is independent of the wavenumber k, there is no dispersion, and the disturbance created at the bottom irregularity travels without change of form. In the more realistic case where $c_p = c_p(k)$, Mork has predicted that waves whose wavenumbers satisfy $c_p(k) = U_{max}$ are most significant.

Cavanie (1969) has developed a model for internal wave generation by the surface tide over the sill at Gibraltar. He has used a hydrostatic model with the Boussinesq approximation and long-wave assumption and has shown the importance of certain nonlinear terms [(such as $v\ (\partial u_x/\partial x)$)] in the formation of internal fronts. This model furnishes a criterion by which the slope of a simple internal wave is increased or decreased according to the conditions upstream of

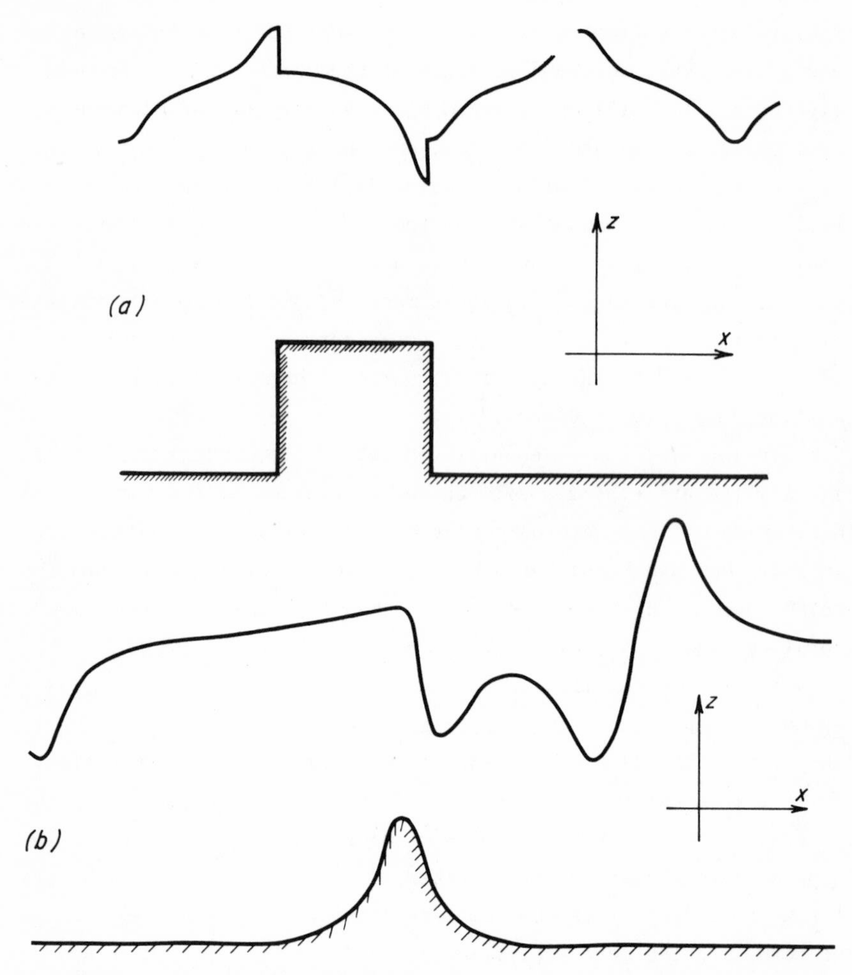

FIG. 2-3. Line of constant density for a first-order internal wave at mid-depth, at the time when the maximum current, U_{max}, is flowing to the right. A linear density profile has been assumed. (a) The case of a square sill with $c_p > U_{max}$. (b) The case for a smooth bump with $c_p < U_{max}$. (Redrawn from Mork, 1968a.)

the wave. In a later paper, Cavanie (1971) has relaxed the long-wave assumption and the Boussinesq approximation. This later model shows the importance of nonlinear terms in the formation of oscillations downstream of the front, a phenomenon often observed east of

Gibraltar. Gade and Ericksen (1969) have also considered the problem of internal wave generation at the Straits of Gibraltar.

Lee (1972) has treated the problem of generation of internal waves by a time-dependent flow over a sill to explain the large amplitude internal waves observed by Halpern (1971a,b) in Massachusetts Bay. In a recent paper, Baines (1973) has discussed the generation of internal tides by flat-bump topography.

The more general problem of lee-wave generation in stratified flows has received much attention. Palm's (1958) introduction to the topic is excellent, as is Miles' (1968) review. More recent work is reported in Davis (1969), Grimshaw (1968), Keady (1971), McIntyre (1972), Miles and Huppert (1969), and Vergeiner (1971).

Continental shelves

The idea, first proposed by Zeilon (1934) that the internal tide may be created when the surface tide (whose currents affect the entire water column) encounters the coast, has been developed by Rattray (1960) for the case of a two-layered ocean. [Ichiye (1963) has modified Rattray's model by considering the surface tide to be a Kelvin wave, so that the surface currents propagate parallel to the coastline.]

When the surface tide encounters a continental shelf, the internal tide is probably created somewhere around the region of A (Fig. 2-4). If the effects of friction are neglected, two internal waves of tidal period are generated: (a) a progressive internal wave which is reflected and travels seaward, and (b) a standing internal wave above the shelf. The amplitude of the reflected wave is less than the amplitude of the shelf wave. The depth -d of the thermocline, as well as the shelf width and the amplitude of the surface tide, determines the amplitudes of the internal tide. If the thermocline is fairly shallow, the amplitudes of the internal tide are less than if the thermocline is deep. If the thermocline lies below the shelf, there is no standing wave. If the shelf is narrow, the amplitudes of the internal tide are small. As the shelf width begins to increase, so do the amplitudes of the generated internal tide. If the

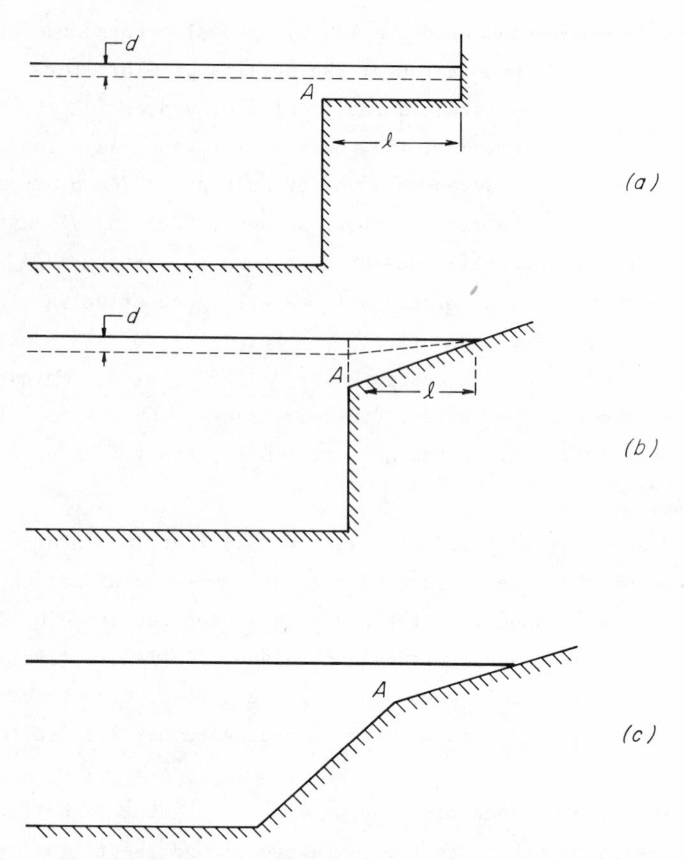

FIG. 2-4. Various shelf models used to explain generation of internal tides. (a) Zeilon (1934); Rattray (1960); Takano and Iida (1969); Weigand *et al.* (1969); Rattray, Dworski, and Kovala (1969), with N constant. (b) Rattray (1960). (c) Prinsenberg (1972: personal communication), with N constant. "A" is a proposed generation region, -d is the depth of the thermocline in a two-fluid system, ℓ is the shelf width.

shelf width becomes sufficiently wide, the amplitudes of the internal tide will again decrease (Takano and Iida, 1969). Shelf width also influences the wavelength of the generated internal tide (Weigand *et al.*, 1969). The amplitudes of the internal tide are directly proportional to the amplitudes of the surface tide. Zalkan (1969) has speculated that there may be a critical amplitude for the surface tide below which no internal tide is generated.

The angle of incidence of the surface tide is unimportant in the generation problem. The situation is shown in Fig. 2-5. Friction acts to decrease the amplitude of the internal tide over the shelf, changing it from a standing wave to a wave which progresses shoreward. This has been demonstrated in a wavetank experiment by Weigand *et al.* (1969).

If the ocean is considered to be continuously stratified, the differential equations may be solved by the method of characteristics (Sec. 4.5), and the paths of the generated waves follow the rays of the reflected surface tide (Fofonoff, 1966; Regal and Wunsch, 1973; Rattray, Dworski, and Kovala, 1969; Larsen *et al.*, 1972). The results of Sec. 4.4 for reflected internal waves may then be applied to the problem.

Straits and channels

The generation of internal tides or internal seiches by interaction of the surface tide with oceanic straits and channels has been dis-

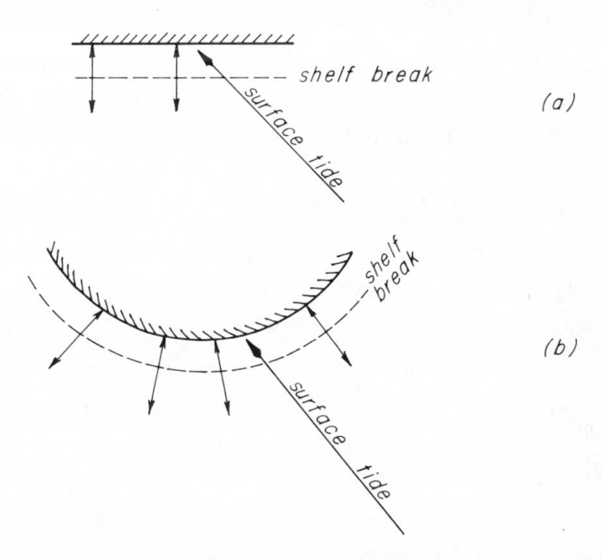

FIG. 2-5. Direction of the internal tide (double-ended arrows) generated by a surface tide whose direction of incidence is not normal to the beach for the case of (a) a straight beach; (b) a curved beach.

cussed theoretically by Niiler (1968) and Gargett (1968), and has been demonstrated in a wavetank experiment by Parrish and Niiler (1971).

2.5 WAKES AND MOVING BODIES

A ship moving slowly through a thin layer of fresh water overlying salty water may produce internal waves on the interface, as has been observed near the coasts (Ekman, 1904; Hughes 1972: personal communication). When this occurs, most of the ship's energy is taken up in generating internal waves, and the vessel seems to be stuck. Defant (1961) has also given an account of this dead water phenomenon, and Lamb (1916), Sretinskii (1959), and Crapper (1967) give a theoretical treatment of the problem.

Internal waves may also be generated by a body moving below the surface of the water. Some wavetank experiments on this problem have already been discussed in Sec. 1.7. Larsen (1969c) has shown that if a neutrally buoyant sphere is slightly displaced from its equilibrium position, it will oscillate, generate internal waves, and lose 90% of its energy in the first two cycles. Hurley (1969) has considered generation of internal waves by vibrating cylinders; as a special case, he allowed the cross-sectional area of the cylinder to vary with time.

When a body, such as a submarine, moves beneath the surface of the water, it may generate lee waves behind it (Lofquist, 1970; Lighthill, 1967; Janowitz, 1968; Mei and Wu, 1964; Hudimac, 1961; Keller and Munk, 1970; Voit and Sebekin, 1969). Miles (1971) has given a review of the problem for a horizontally moving source.

If the fluid is continuously stratified or multilayered, the collapsing of the submarine's wake is a much more efficient generator of internal waves than the movement of the submarine *per se* (Schooley and Stewart, 1963), except at low speeds (Bell, 1973b). The idea is that a cylindrical submarine leaves behind it a fairly cylindrical and well-mixed wake. As the fluid inside the wake seeks its equilibrium level, it flattens and deforms the isopycnals, which then oscillate (Fig. 2-6). Studies on the amplitudes and phases of internal

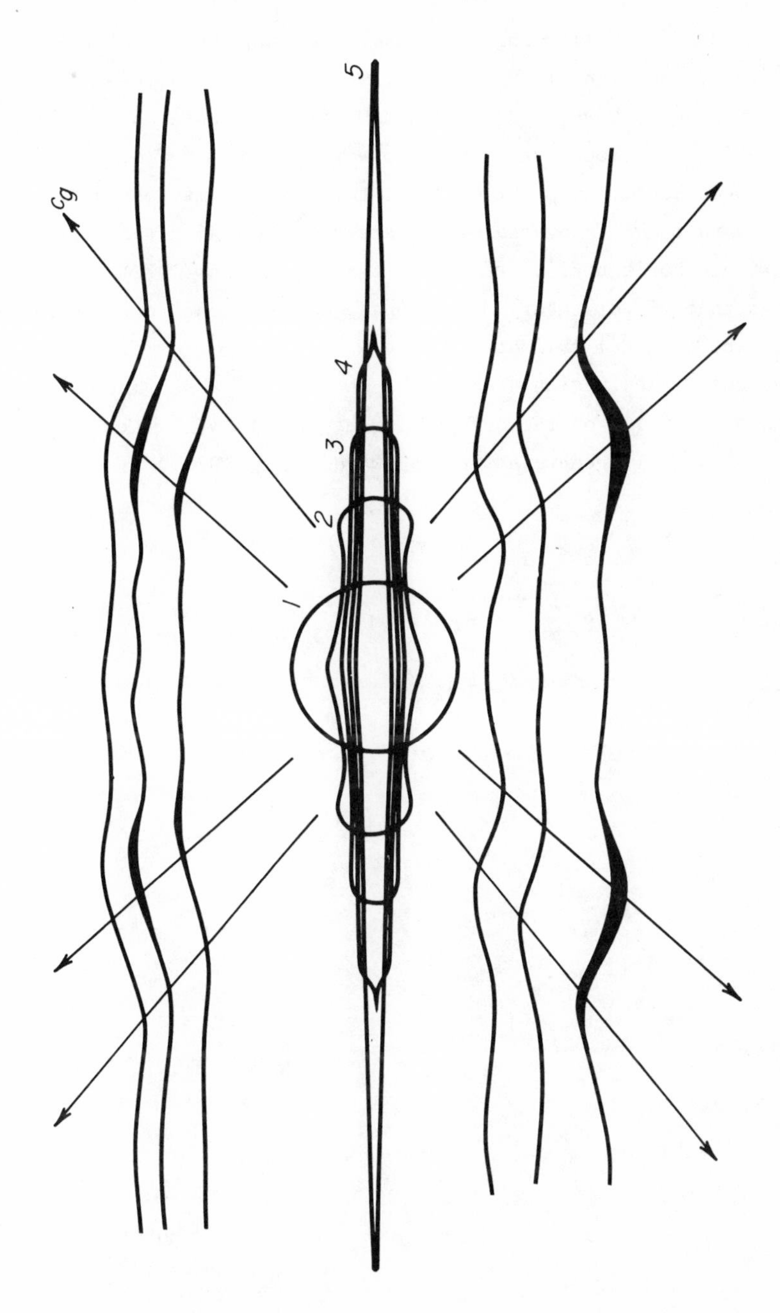

FIG. 2-6. Profiles of a collapsing wake at various stages. The internal wave pattern of the isopycnals is shown for stage 3. Stages 1 and 2 are the initial collapse stage. Stages from 3 to 5 are the principal collapse stage. The final collapse stage occurs after stage 5. Arrows point in the direction of group velocity, and their shafts are lines of constant phase. (Adapted from Wu, 1969.)

waves generated by collapsing wakes have been done by Schooley and Hughes (1972) and by Wu (1969). Mei (1968) has given a theoretical discussion of Wu's work, and Young and Hirt (1972) have done a numerical model of it. Young and Hirt, who worked directly from the Navier-Stokes equation, point out that their method may be used to describe a variety of internal wave patterns, including the effects of a variable horizontal current. Hartman and Lewis (1972) have cautioned that linear theory may be inadequate for the region very close to the original cylindrical wake.

The collapsing region need not be the wake of a submarine. Currents over a sill on the ocean floor may create well-mixed regions whose collapse will generate internal waves (Lee, 1972).

CHAPTER 3

THEORY OF PROPAGATING AND STANDING INTERNAL WAVES OVER A HORIZONTAL OCEAN FLOOR

3.1 INTRODUCTION

Internal waves can be classified according to the density distribution of the fluid in which they propagate (Bretherton, 1971): (a) interfacial waves, the sort that occur at the interface of a two-fluid system; (b) plane waves, which occur when the density of the fluid increases linearly with depth; and (c) waves of a mixed type, which include internal waves occurring in a fluid whose density varies continuously, but not necessarily linearly. Different classes of internal waves have different physical properties. For instance, the group and phase velocities of interfacial waves point in the same direction (like the group and phase velocities of surface waves), while for plane waves they are almost at right angles (Sec. 3.4). Internal waves in the ocean are of a mixed type; mathematical feasibility, however, often forces the investigator to classify them as either interfacial or plane waves, depending on the particular characteristic he is examining. [Mathematically, a continuous density can be considered as the limit of n layers, as n becomes large (Burnside, 1889); likewise, a layered system can be considered as the limit of a continuous density (Yih, 1960). Usually, however, there is a distinction made between continuous and discontinuous density distributions.] Roberts (1973) has keyworded most of the papers on internal waves in the ocean which appeared before 1972, including a keyword or keywords for the density distributions which the paper assumes. Section 3.4 discusses solutions of the internal wave equation for these various density distributions.

While the generating mechanisms for internal waves are not well-defined, and the dissipation rates for internal waves are not known, the problem of a small amplitude internal wave propagating linearly over a horizontal bottom is well-posed, and theory agrees remarkably well with observation. Our review of the theory begins with an example from Tolstoy (1963). If we dye a particle of water inside a fluid and displace it slightly, the particle will oscillate around its equilibrium position, as a mass on a spring does in simple harmonic motion. If the fluid has a sharp increase in density somewhere (as in an oceanic thermocline), and if this region is slightly perturbed, internal waves will result. The water on either side of the thermocline moves also, but much less than the water in the region of sharper increase of density. The surface of the water hardly moves at all.

To be specific, consider a vertically stratified compressible fluid (in geophysical parlance, a fluid being stratified says nothing about the form of the density profile except that it is not constant) at rest in a gravitational field. We use a stationary Cartesian coordinate system, x, y, z, with z the vertical axis pointing up and $\vec{g}$, the gravity field, pointing down (Fig. 3-1).

Let ζ be the z-component of displacement of a fluid particle from its equilibrium position, and $\bar{\rho}(z)$ the equilibrium density of the fluid. Consider a single particle of fluid displaced vertically from its equilibrium position, independently of its neighbors. When we release it, its motion will obey the equation

$$\bar{\rho} \frac{d^2\zeta}{dt^2} = - g\Delta\bar{\rho} \tag{1}$$

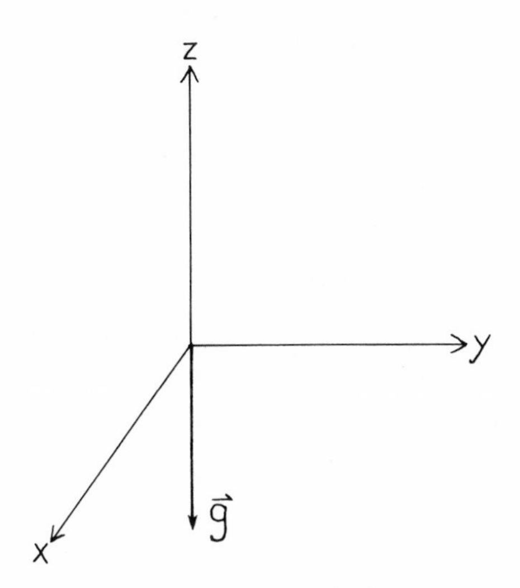

FIG. 3-1. Geometry of the system.

where $\Delta\bar{\rho}$ is composed of two terms

$$\Delta\bar{\rho} = \Delta\bar{\rho}_1 + \Delta\bar{\rho}_2 \tag{2}$$

The first term ignores compressibility and is given by

$$\Delta\bar{\rho}_1 = -\zeta \frac{d\bar{\rho}}{dz} \tag{3}$$

The other term, $\Delta\bar{\rho}$, is the change in density due to compressibility, which, in this case, is produced by the change in hydrostatic head, or pressure p

$$\Delta p = \zeta \frac{dp}{dz} \tag{4}$$

But we know that for adiabatic expansion (e.g., Eskinazi, 1968, p. 331)

$$\Delta\bar{\rho}_2 = \frac{1}{V_s^2} \Delta p \tag{5}$$

where V_s is the speed of sound in the fluid.

Since the fundamental equation of hydrostatics is

$$\frac{dp}{dz} = -\bar{\rho}g \tag{6}$$

we have

$$\Delta\bar{\rho}_2 = \frac{1}{V_s^2}\zeta(-\bar{\rho}g) \tag{7}$$

and Eqs. (1) and (2) give

$$\bar{\rho}\frac{d^2\zeta}{dt^2} = g\left(\zeta\frac{d\bar{\rho}}{dz} + \frac{g}{V_s^2}\bar{\rho}\zeta\right)$$

or

$$\frac{d^2\zeta}{dt^2} - g\left(\frac{1}{\bar{\rho}}\frac{d\bar{\rho}}{dz} + \frac{g}{V_s^2}\right)\zeta = 0 \tag{8}$$

When the term in parentheses is negative, the fluid particle goes into simple harmonic motion such that the square of its angular frequency is given by

$$N^2 = -g\left(\frac{1}{\bar{\rho}}\frac{d\bar{\rho}}{dz} + \frac{g}{V_s^2}\right) \tag{9}$$

The manner in which we have chosen our coordinate system implies that $d\bar{\rho}/dz < 0$. Thus, for a sufficiently pronounced density variation

$$\left|\frac{1}{\bar{\rho}}\frac{d\bar{\rho}}{dz}\right| > \frac{g}{V_s^2}$$

the situation is stable, and the medium is characterized for each value of z by a resonant frequency of oscillation, N(z). N is defined by Eq. (9) and is variously called the Väisälä, Brunt, Brunt-Väisälä, intrinsic, buoyancy, stability, or limiting frequency. Defant (1961) has used the quantity $E \equiv N^2/g$ and called it the stability. Krauss (1966b) has used $\Gamma = \frac{1}{\bar{\rho}}\frac{d\bar{\rho}}{dz}$ (his z-axis points downward), and Miles (1961) and others have used $\beta(z) = -\frac{1}{\bar{\rho}(z)}\frac{d\bar{\rho}}{dz}$. In this book we follow the most standard oceanographic terminology, use N, and call it the "Brunt-Väisälä frequency".

3.2 THE BASIC EQUATIONS

The main purpose of this section is to develop in detail a general equation for the vertical velocity w^* [the asterisks denote quantities which will soon be broken up into a mean term and a perburbation term, see Eq. (6); they do not denote complex conjugates here], and to reconcile the notation in the equations of two basic references on internal waves, Krauss (1966b) and Phillips (1966). Readers who are familiar with the development of the vertical velocity equation may wish to skip this section.

We assume a Cartesian coordinate system x, y, z (z positive upward) with unit vectors $\hat{i}$, $\hat{j}$, $\hat{k}$. The x-, y-, and z-components of the instantaneous velocity are u^*, v^*, and w^*, respectively, so the velocity vector $\vec{u}^*$ may be written

$$\vec{u}^* = \hat{i}u_x^* + \hat{j}v^* + \hat{k}w^*$$

where the scalars u_x^*, v^*, and w^* are functions of x, y, z, and t. The density and pressure are given by $\rho^* = \rho^*(x,y,z,t)$, and $p^* = p^*(x,y,z,t)$, respectively.

We begin with two equations. The first is the equation of continuity

$$\frac{D\rho^*}{Dt} + \rho^*(\nabla \cdot u^*) = 0 \qquad (1)$$

where

$$\frac{D}{Dt} \equiv \frac{\partial}{\partial t} + u_x^* \frac{\partial}{\partial x} + v^* \frac{\partial}{\partial y} + w^* \frac{\partial}{\partial z}$$

(For a clear discussion of D/Dt, see Bird, Stewart, and Lightfoot, 1960, p. 73). The second equation is the Navier-Stokes equation with Coriolis force (for an explanation of how the Coriolis term arises in the equation of motion, see Constant, 1954, p. 69-73), and an arbitrary force $\vec{F} = \hat{i}F_x + \hat{j}F_y + \hat{k}F_z$ added (Batchelor, 1970, p. 147)

$$\rho^* \frac{D\vec{u}^*}{Dt} + 2\rho^*(\vec{\Omega} \times \vec{u}^*) + \nabla p^* + \hat{k}g\rho^*$$
$$= \mu\nabla^2\vec{u}^* + \frac{\mu}{3}\nabla(\nabla\cdot\vec{u}^*) + \vec{F} \quad (2)$$

Here

$$\vec{\Omega} = \hat{i}\,\Omega_x + \hat{j}\,\Omega_y + \hat{k}\,\Omega_z \quad (3)$$

is the angular velocity of the earth, and μ is the coefficient of viscosity, taken as constant.

For incompressible fluids, neglecting diffusion,

$$\frac{D\rho^*}{Dt} = 0 \quad (4)$$

The physical consequence of this assumption is that compression, or sound waves are excluded from consideration. This is quite reasonable because sound waves play little role in the overall dynamics of the upper ocean (Phillips, 1966, p. 15). Clearly, ρ^* is never 0, and Eq. (1) now becomes

$$\nabla \cdot \vec{u}^* = 0 \quad (5)$$

If the density ρ^* is the sum of a mean, $\bar{\rho}(z)$, and a fluctuation $\rho(x,y,z,t)$ resulting from fluid motion, then

$$\rho^* = \bar{\rho}(z) + \rho(x,y,z,t) \quad (6)$$

Similarly, let

$$p^* = \bar{p}(z) + p(x,y,z,t) \quad (7)$$

and

$$u_x^* = U_x\,(x,y,z) + u_x(x,y,z,t) \quad (8a)$$

$$v^* = V(x,y,z) + v(x,y,z,t) \quad (8b)$$

$$w^* = W(x,y,z) + w(x,y,z,t) \quad (8c)$$

Thus,

$$\vec{u}^* = \hat{i}u_x^* + \hat{j}v^* + \hat{k}w^* = \hat{i}(U_x + u_x) + \hat{j}(V + v) + \hat{k}(W + w)$$
$$= \vec{U} + \vec{u} \quad (8d)$$

Here, $\vec{U}$ describes the time-independent flow, while $\vec{u}$ describes the time-dependent fluctuations of that flow. Eqs. (2), (5), (6), and (7) may be combined to give

$$(\bar{\rho} + \rho)\frac{D\vec{u}^*}{Dt} + 2(\bar{\rho} + \rho)\vec{\Omega} \times \vec{u}^* + \nabla(\bar{p} + p) + \hat{k}g(\bar{\rho} + \rho) = \mu \nabla^2\vec{u}^* + \vec{F} \tag{9}$$

In the absence of motion and outside forces, the vertical component of Eq. (9) is

$$\frac{\partial \bar{p}}{\partial z} + g\bar{\rho} = 0 \tag{10}$$

which is just the equation of hydrostatics.

<u>If the fluctuation in density, ρ, is neglected in all the terms except $\hat{k}g(\bar{\rho} + \rho)$</u>, then Eq. (9) becomes

$$\bar{\rho}\frac{D\vec{u}^*}{Dt} + 2\bar{\rho}(\vec{\Omega} \times \vec{u}^*) + \nabla(\bar{p} + p) + \hat{k}g(\bar{\rho} + \rho) = \mu \nabla^2\vec{u}^* + \vec{F} \tag{11}$$

Since $\bar{p} = \bar{p}(z)$, this last equation may be rewritten by using Eq. (10) and dividing through by $\bar{\rho}$,

$$\frac{D(\vec{U} + \vec{u})}{Dt} + 2\vec{\Omega} \times (\vec{U} + \vec{u}) + \frac{1}{\bar{\rho}}\nabla p + \frac{\hat{k}g\rho}{\bar{\rho}} = \frac{\mu}{\bar{\rho}}\nabla^2(\vec{U} + \vec{u}) + \frac{\vec{F}}{\bar{\rho}} \tag{12}$$

Eq. (12) is equivalent to Phillips' (1966) Eq. (2.4.7); a proof of this given in Appendix 3.2.1.

Assuming that terms which cause Eq. (12) to be nonlinear may be disregarded (the usual rationale for neglecting, for example, $u_x(\partial u_x/\partial x)$, is that u_x is small and $\partial u_x/\partial x$ is small, so their product is smaller yet), and taking

$$\vec{U} = \vec{F} = 0, \quad \mu = \Omega_x = \Omega_y = 0, \text{ and } 2\Omega_z \equiv f$$

Eqs. (12), (4), and (5) become

$$\frac{\partial u_x}{\partial t} - fv + \frac{1}{\bar{\rho}}\frac{\partial p}{\partial x} = 0 \tag{13a}$$

$$\frac{\partial v}{\partial t} + fu_x + \frac{1}{\bar{\rho}}\frac{\partial p}{\partial y} = 0 \tag{13b}$$

$$\frac{\partial w}{\partial t} + \frac{1}{\bar{\rho}}\frac{\partial p}{\partial z} + \frac{g\rho}{\bar{\rho}} = 0 \tag{13c}$$

$$\frac{\partial \rho}{\partial t} + w\frac{\partial \bar{\rho}}{\partial z} = 0 \tag{13d}$$

$$\frac{\partial u_x}{\partial x} + \frac{\partial v}{\partial y} + \frac{\partial w}{\partial z} = 0 \tag{13e}$$

These important equations appear in Lamb (1945, p. 378).

The assumption that ρ may be neglected in all terms except the gravitational term is a simplified form of the *Boussinesq approximation* (Boussinesq, 1903, cited by Spiegel and Veronis, 1960, who give a discussion of the Boussinesq approximation for a compressible fluid). The Boussinesq approximation has been subject to some scrutiny (Groen, 1948a; Long, 1965; Benjamin, 1966, 1967; Thorpe, 1968c). However, it is usually valid, except for some cases of solitary waves. For this reason, we follow common practice and assume that it holds as we develop equations for internal gravity waves.

After some algebraic manipulation, Eq. (12) may be written in terms of the vertical velocity w^*, where $w^* = W + w$. Details of the computation are given in Appendix 3.2.2. The vertical velocity equation is:

$$\frac{\partial}{\partial t^2}\nabla^2 w^* + N^2\nabla_h^2 w^* + 4\frac{N^2}{g}\vec{\Omega}\cdot[\nabla \times (\vec{\Omega} \times \vec{u}^*)]$$
$$- 4\vec{\Omega}\cdot \nabla \times \vec{\Omega} \times \frac{\partial \vec{u}^*}{\partial z} - \frac{N^2}{g}\frac{\partial^3 w^*}{\partial z \partial t^2} = Q \tag{14}$$

Q contains nonlinear terms and is given by

$$Q = \frac{1}{\bar{\rho}}\frac{\partial}{\partial z}\Bigg(2\vec{\Omega}\cdot\Bigg\{\nabla \times \hat{k}g\rho - \nabla \times [\mu\nabla^2\vec{u}^* - \bar{\rho}(\vec{u}^*\cdot\nabla)\vec{u}^* + \vec{F}]$$
$$+ \frac{d\bar{\rho}}{dz}\left[\hat{i}\left(\Omega_x w^* - \Omega_z u_x{}^* - \frac{\partial v}{\partial t}\right) + \hat{j}\left(\Omega_y w^* - \Omega_z v^* + \frac{\partial u_x^*}{\partial t}\right)\right]\Bigg\}$$
$$- 2\bar{\rho}\frac{\partial^2}{\partial z \partial t}(\Omega_x v^* - \Omega_y u_x^*) - \nabla_h\cdot\frac{\partial}{\partial t}\mu\nabla^2\vec{u}^* - [\bar{\rho}(\vec{u}^*\cdot\nabla)\vec{u}^* + \vec{F}]\Bigg)$$

$$+ \frac{1}{\bar{\rho}} \nabla_h^2 \left(g(\vec{u}^* \cdot \nabla \rho) + \frac{\partial}{\partial t} [\mu \nabla^2 \vec{u}^* - \bar{\rho}(\vec{u}^* \cdot \nabla)\vec{u}^* + \vec{F}]_z - 2\bar{\rho} \frac{\partial}{\partial t} (\Omega_x v^* - \Omega_y u_x^*) \right) \tag{15}$$

The Brunt-Väisälä frequency N is given by

$$N^2 = N^2(z) \equiv - \frac{g}{\bar{\rho}} \frac{d\bar{\rho}}{dz} \tag{16}$$

Equation (14) contains the vertical velocity equations of Krauss (1966b, p. 8) and Phillips (1966, p. 191), as we now show.

If we assume

$$2\vec{\Omega} = \hat{k}2\Omega_z = \hat{k}f; \quad \vec{U} \equiv 0; \quad \mu = 0 \tag{17}$$

and if $(\vec{u}^* \cdot \nabla)\vec{u}^*$ and $(\vec{u}^* \cdot \nabla\rho)$ are neglected, then Eq. (14) becomes

$$\frac{\partial^2}{\partial t^2} \nabla^2 w + N^2 \nabla_h^2 w + f^2 \frac{\partial^2 w}{\partial z^2} - \frac{N^2}{g} \left(\frac{\partial^3 w}{\partial z \partial t^2} + f^2 \frac{\partial w}{\partial z} \right) = - \frac{1}{\bar{\rho}} \left[f \left(\nabla \frac{\partial \vec{F}}{\partial z} \right)_z + \nabla_h \cdot \frac{\partial^2 \vec{F}}{\partial z \partial t} - \nabla_h^2 \frac{\partial F_z}{\partial t} \right] \tag{18}$$

which is Krauss' Eq. (112.10). Krauss takes his z-axis positive downward, so that $d\bar{\rho}/dz > 0$; this explains the sign difference for the fourth term on the left-hand side of Eq. (18).

With

$$Q \equiv 0 \quad \text{and} \quad \vec{U} \equiv 0 \tag{19}$$

and assuming $\bar{\rho}$ may be replaced by a constant, ρ_0 in the inertia terms [this assumption will make the third and fifth terms on the left-hand side of Eq. (14) disappear; see Appendix 3.2.2], we have

$$\frac{\partial^2}{\partial t^2} \nabla^2 w + N^2 \nabla_h^2 w - 4\vec{\Omega} \cdot \left[\nabla \times \left(\vec{\Omega} \times \frac{\partial \vec{u}}{\partial z} \right) \right] = 0$$

If we define vorticity by

$$\nabla \times \vec{u} = \hat{i} \left(\frac{\partial w}{\partial y} - \frac{\partial v}{\partial z} \right) + \hat{j} \left(\frac{\partial u_x}{\partial z} - \frac{\partial w}{\partial x} \right) + \hat{k} \left(\frac{\partial v}{\partial x} - \frac{\partial u_x}{\partial y} \right) \tag{20}$$

and assume that $\partial w/\partial y = \partial v/\partial z$ and $\partial u/\partial z = \partial w/\partial x$, it is shown in Appendix 3.2.3 that

$$- 4\vec{\Omega}\cdot\left[\nabla \times \left(\vec{\Omega} \times \frac{\partial \vec{u}}{\partial z}\right)\right] = (2\vec{\Omega}\cdot\nabla)^2 w \tag{21}$$

so that Eq. (14) finally becomes

$$\frac{\partial^2}{\partial t^2} \nabla^2 w + [(2\vec{\Omega}\cdot\nabla)^2 + N^2 \nabla_h^2]w = 0 \tag{22}$$

which is Phillips' Eq. (5.7.3). The reader may verify that the right-hand side of Eq. (14) contains Philipps' Eq. (5.2.5).

If we assume, as above, that $2\vec{\Omega} = \hat{k}f$, then Eq. (22) becomes

$$\frac{\partial^2}{\partial t^2} \nabla^2 w + f^2 \frac{\partial^2 w}{\partial z^2} + N^2 \nabla_h^2 w = 0 \tag{23}$$

All of this is much easier if we want only the linearized vertical velocity equation with the Boussinesq approximation, Eq. (23). In that case, we can proceed directly from Eq. (13) (see Appendix 3.2.4).

Sometimes Eq. (23) is written in terms of the streamfunction $\psi(x,z,t)$ defined by

$$w = - \frac{\partial \psi}{\partial x} \qquad u = \frac{\partial \psi}{\partial z} \tag{24}$$

In that case, Eq. (23) becomes

$$\frac{\partial}{\partial x}\left(\frac{\partial^2}{\partial t^2} \nabla^2 \psi + f^2 \frac{\partial^2 \psi}{\partial z^2} + N^2 \frac{\partial^2 \psi}{\partial x^2}\right) = 0$$

one solution of which is

$$\frac{\partial^2}{\partial t^2} \nabla^2 \psi + f^2 \frac{\partial^2 \psi}{\partial z^2} + N^2 \frac{\partial^2 \psi}{\partial x^2} = 0 \tag{25}$$

(For instance, see Magaard, 1965; Wunsch, 1968; Yih, 1960; McEwan, 1971.)

No general solution has been found for Eq. (14), but various approximations have been made so that it can be solved. These solutions will be discussed in Sec. 3.4.

3.3 FREQUENCY BOUNDS, f AND N, FOR INTERNAL GRAVITY WAVES

We begin with Eq. (3.2.18), which is an approximation to the vertical velocity Eq. (3.2.14), and assume that the vertical velocity w is periodic in time with period ω :

$$w = \hat{w}(x,y,z)\exp(-i\omega t) \tag{1}$$

If ω is constant, and if $F \equiv 0$, Eq. (3.2.18) becomes

$$\nabla_h^2 \hat{w} - \frac{\omega^2 - f^2}{N^2 - \omega^2}\left(\frac{\partial^2 \hat{w}}{\partial z^2} - \frac{N^2}{g}\frac{\partial \hat{w}}{\partial z}\right) = 0 \tag{2}$$

The coefficient

$$\frac{\omega^2 - f^2}{N^2 - \omega^2} \tag{3}$$

determines what sort of partial differential equation Eq. (2) will be. If

$$\frac{\omega^2 - f^2}{N^2 - \omega^2} < 0$$

then Eq. (2) is called an elliptical partial differential equation and, in the absence of a free surface or interfaces, no wave motion is possible (Yih, 1969). If

$$\frac{\omega^2 - f^2}{N^2 - \omega^2} > 0 \tag{4}$$

Eq. (2) is called hyperbolic, and wave motion is possible. If $N^2(z) > f^2$ for all z, i.e., at all depths, Eq. (4) is equivalent to

$$f^2 < \omega^2 < N^2(z) \tag{5}$$

Thus the frequency of internal gravity waves at a given depth is bounded above by the Brunt-Väisälä frequency N(z), a fact first noted by Groen (1948a,b), and below by the inertial frequency f. N(z) may be as large as 0.042 sec^{-1} (period T_N = 150 sec = 2.5 min). The inertial frequency f varies from about 1.4×10^{-4} sec^{-1} at the poles ($T_f \doteq 12$ hr) to 0 at the equator (T_f infinite). We therefore have Fig. 3-2 representing allowable frequencies (periods) for internal gravity waves in the ocean.

At a given depth z, the Brunt-Väisälä frequency N(z) determines the maximal frequency for waves which occur at that depth. It follows that, since N(z) is greatest in the thermocline where the variation of density with depth is greatest, internal waves of largest frequency (and hence shortest period) occur in the thermocline.

Observations indicate that N is actually a remarkably sharp cutoff frequency for internal gravity waves. In examining internal waves which were generated in a wavetank by a small cylinder oscillating with frequency ω, Mowbray and Rarity (1967a) have found that when $\omega = 1.11$ N, "no wave-like disturbances are observed in the steady state, disturbances being confined to a mixing region close to the cylinder." The spectra of many oceanic observations of internal waves likewise show a cutoff at N (Garrett and Munk, 1972b). Phillips (1971) points out that if the density is layered, sampling at a fixed depth may yield data which extend the spectrum above the Brunt-Väisälä frequency, which no doubt is one reason that the observed cutoff at N is not sharper. In addition, the limits N and f may be affected by currents in the ocean (Saint-Guily, 1970; Magaard, 1968; Bell, 1974a), and by wind blowing across the ocean (Tomczak, 1966).

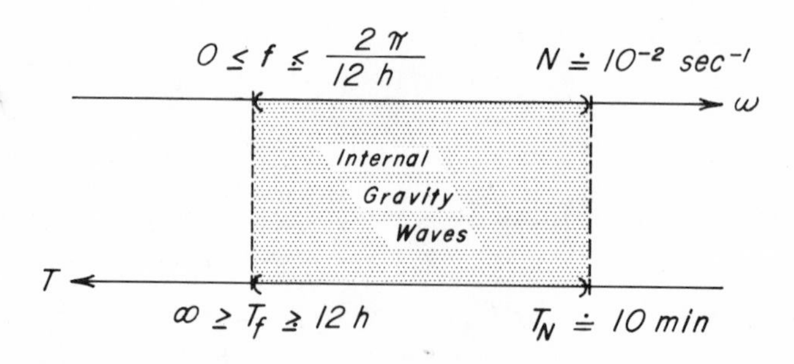

FIG. 3-2. Allowable frequencies (periods) for internal gravity waves.

We have defined N as

$$\left[-\frac{g}{\bar{\rho}(z)}\frac{d\bar{\rho}}{dz}\right]^{1/2} \tag{6}$$

In the literature, the definition $\left(-\frac{g}{\rho_0}\frac{d\bar{\rho}}{dz}\right)^{1/2}$, where $\rho_0 = \bar{\rho}(0) =$ constant, is also used (for instance, Phillips, 1966, p. 17). If we suppose that the density has an exponential profile, i.e., $\bar{\rho}(z) = \rho_0 \exp(-N^2 z/g)$, then $\left(-\frac{g}{\bar{\rho}}\frac{d\bar{\rho}}{dz}\right) = N^2$ is constant; while if we consider the density to be linear, i.e., $\bar{\rho}(z) = \rho_0[c_1 - (N^2 z/g)]$ (c_1 an arbitrary constant), then $\left(-\frac{g}{\rho_0}\frac{d\bar{\rho}}{dz}\right) = N^2$ is constant. As may be seen from the table below, the two definitions give very similar results. Indeed, the first two terms in the Maclaurin's series for $\rho_0 \exp(-N^2 z/g)$ are $\rho_0 + z[\rho_0(-N^2/g)] = \rho_0[1-(N^2 z/g)]$, so that the linear density profile is a first-order approximation to the exponential density profile.

3.4 EIGENSOLUTIONS OF THE VERTICAL VELOCITY EQUATION FOR VARIOUS DENSITY DISTRIBUTIONS

Solutions of the vertical velocity equation for internal waves, Eq. (3.2.14), have been found after various simplifications are made. Most solutions require that the ocean floor be horizontal and impermeable; i.e., that

$$w = 0 \quad \text{at} \quad z = -D \tag{1}$$

where -D is the depth of the ocean. The cases of variations in bottom topography are discussed in Chap. 4. We also assume that the upper surface is fixed; that is

$$w = 0 \quad \text{at} \quad z = 0 \tag{2}$$

Phillips (1966, p. 165) has shown that this is usually the case for internal waves, and Yih (1960) has discussed the mathematical aspects of the assumption. More exact boundary conditions at the surface and at the bottom are given in Krauss (1966b, pp. 9-12).

TABLE 3-1

Density computed from (a) $\bar{\rho} = \rho_0(c_1 - N^2 z/g)$ and (b) $\bar{\rho} = \rho_0 \exp(-N^2 z/g)$; with the values $\rho_0 = 1$, $c_1 = 1$, and $N^2/g = 10^{-5} m^{-1}$ (T_N = 10 min).

-z (m)	(a) Linear density	(b) Exponential density
100	1.00100	1.00100
200	1.00200	1.00200
300	1.00300	1.00300
400	1.00400	1.00400
500	1.00500	1.00501
1000	1.01000	1.01005
2000	1.02000	1.02020
3000	1.03000	1.03045
4000	1.04000	1.04081

As in Sec. 3.3, we again assume that no external forces are acting, i.e., that $F \equiv 0$, and that we have free waves of the form

$$w = \hat{w}(x,y,z) \exp(i\omega t) \tag{3}$$

Taking boundary conditions (1) and (2), and supposing that density and its variation with depth depend only on depth, so that $N = N(z)$, Eq. (3) is separable

$$\hat{w}(x,y,z) = \mathcal{W}(z)F(x,y) \tag{4}$$

and Eq. (3) becomes

$$\frac{1}{\omega^2 - f^2}\frac{\nabla_h^2 F}{F} - \frac{1}{N^2 - \omega^2}\left(\frac{d^2\mathcal{W}}{dz^2} - \frac{N^2}{g}\frac{d\mathcal{W}}{dz}\right)\frac{1}{\mathcal{W}} = 0 \tag{5}$$

The first term in this equation depends only on x and y, and the second term depends only on z. The only way this can be true for all x, y, and z is to have

$$\frac{1}{\omega^2 - f^2}\frac{\nabla_h^2 F}{F} = \frac{1}{N^2 - \omega^2}\left(\frac{d^2\mathcal{W}}{dz^2} - \frac{N^2}{g}\frac{d\mathcal{W}}{dz}\right)\frac{1}{\mathcal{W}} = -\nu$$

where ν is a separation constant. Then

$$\nabla_h^2 F + (\omega^2 - f^2)\nu F = 0 \tag{6a}$$

and

$$\frac{d^2\mathcal{W}}{dz^2} - \frac{N^2}{g}\frac{d\mathcal{W}}{dz} + (N^2 - \omega^2)\nu\mathcal{W} = 0 \tag{6b}$$

Equation (6a) describes the waveform of an internal wave, and we return to it in Sec. 3.5. Equation (6b) is a vertical velocity equation for internal waves and is sometimes called the structure equation (Chapman and Lindzen, 1970, p. 116).

Following Krauss (1966b, p. 20) we set

$$\nu = \frac{k_h^{\ 2}}{\omega^2 - f^2} \tag{7}$$

where k_h is a horizontal wavenumber

$$k_h^{\ 2} = k_x^{\ 2} + k_y^{\ 2}$$

Also let

$$q(z) = [N^2(z) - \omega^2]\bar{\rho}(z)$$

Rewriting Eq. (6b) along with its boundary conditions (1) and (2) using Eqs. (3) and (4), we have

$$\frac{d}{dz}\left[\bar{\rho}(z)\frac{d\mathcal{W}}{dz}\right] + \left[\frac{N^2(z) - \omega^2}{\omega^2 - f^2}\right]k_h^{\ 2}\,\bar{\rho}(z)\,\mathcal{W} = 0 \qquad \text{Sturm-Liouville Equation}$$

or

$$\frac{d}{dz}\left[\bar{\rho}(z)\frac{d\mathcal{W}}{dz}\right] + \nu q(z)\,\mathcal{W} = 0 \tag{8a}$$

$$\mathcal{W}(z) = 0, \qquad z = -D \tag{8b}$$

$$\mathcal{W}(z) = 0, \qquad z = 0 \tag{8c}$$

For $\nu q(z) > 0$, we thus have an eigenvalue problem: Each choice for the eigenvalue ν_n will yield an eigenfunction solution $\mathcal{W}_n(z)$ for Eq. (8a). That is, for a fixed f and fixed ω, solutions $\mathcal{W}_n(z)$ to Eq. (8a) can be found for an infinity of horizontal wavenumbers $k_h = k_{h_0}, k_{h_1}, k_{h_2}, \ldots, k_{h_n}, \ldots$ The internal wave characterized by $\mathcal{W}_n$ is called an nth-order internal wave, or an internal wave of the nth mode. These solutions are conveniently ordered according to their number of extrema between the surface and the bottom: $\mathcal{W}_1$ has one extremum, $\mathcal{W}_2$ has two extrema, and so on (Fig. 3-3). The case n = 0 corresponds to a surface wave solution. Although we have excluded this case from consideration here, it may be included by specifying a more exact boundary condition at the surface (Krauss, 1966b, pp. 9-12).

Remember that in order to have wave solutions to Eq. (8a), ω must be less than N(z). This need not hold for all z; for a given ω, it is possible to consider internal wave solutions only in that region where $\omega < N(z)$ does hold (Phillips, 1966, pp. 164-165; Krauss, 1966b, pp. 26-29).

The vertical displacement $\zeta(x,y,z,t)$ is directly proportional to the eigenfunction $\mathcal{W}(z)$. This is most easily seen if we assume that an internal wave has the form $\exp[i(\vec{k}_h \cdot \vec{r} - \omega t)]$. Then $w = \mathcal{W}(z)\exp[i(\vec{k}_h \cdot \vec{r} - \omega t)]$ and $\zeta = a(z)\ \exp[i(\vec{k}_h \cdot \vec{r} - \omega t)]$. By definition $w = d\zeta/dt$. If $d\zeta/dt \doteq \partial\zeta/\partial t$, then at a given depth $z = z_o$ we have $\mathcal{W}(z_o) \doteq -i\omega a(z_o)$, so that

$$\left|\zeta(x,y,z_o,t)\right| = \left|a(z_o)\right| \doteq \left|\frac{1}{\omega}\right|\left|\mathcal{W}(z_o)\right|$$

Thus, once $\mathcal{W}_n$ is determined, we have an idea of the relative vertical displacement of the isopycnals for the nth internal wave mode. The vertical displacement of the isopycnals due to internal waves of the first five modes is shown in Fig. 3-4.

Because of the linearity of Eq. (8a), arbitrary sums of $\mathcal{W}_n$ are also solutions of Eq. (8a).

If we consider two-dimensional waves, so that $\vec{k} = \hat{i}k_x + \hat{j}k_y = \vec{k}_h$, and if we neglect the earth's rotation, then for an nth-order internal wave, Eq. (7) becomes

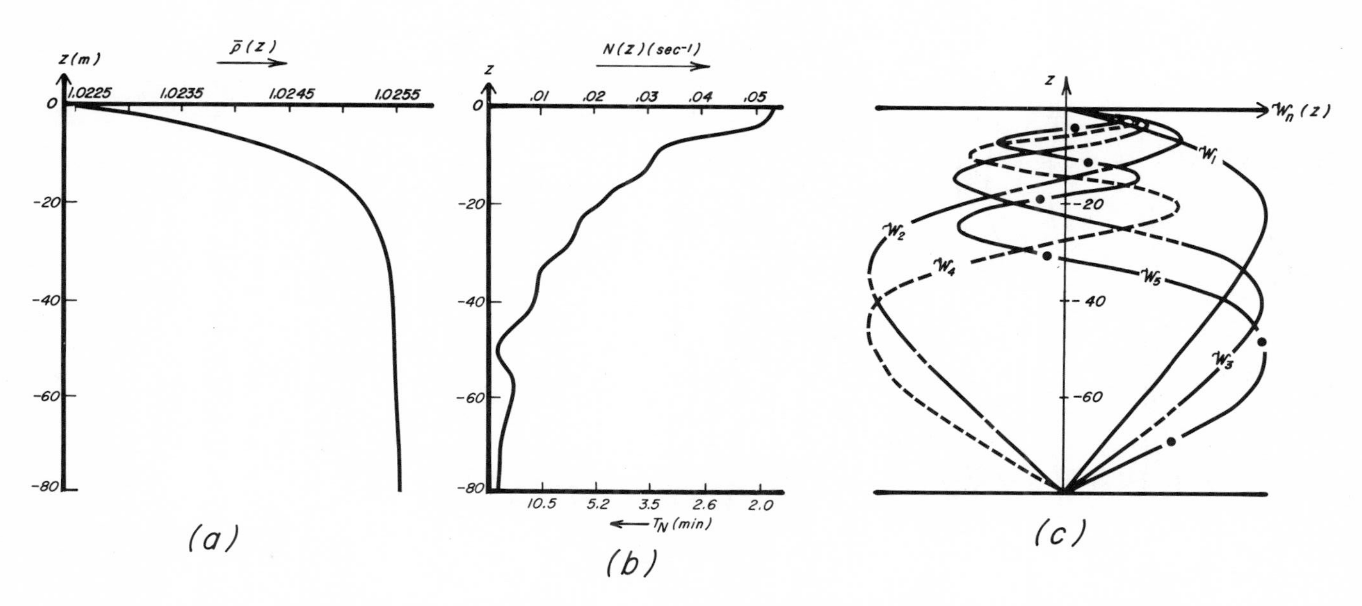

FIG. 3-3. (a) Density data from Massachusetts Bay used in determining (b) N(z). (c) Relative amplitudes of the eigensolution $\mathcal{W}_n(z)$ (n = 1,2,3,4,5) of Eq. (10a). These particular eigensolutions were determined numerically using $\bar{\rho}_i(z)$ (i = 1,...,80). (Redrawn from Halpern, 1971a,b; used with the author's permission.)

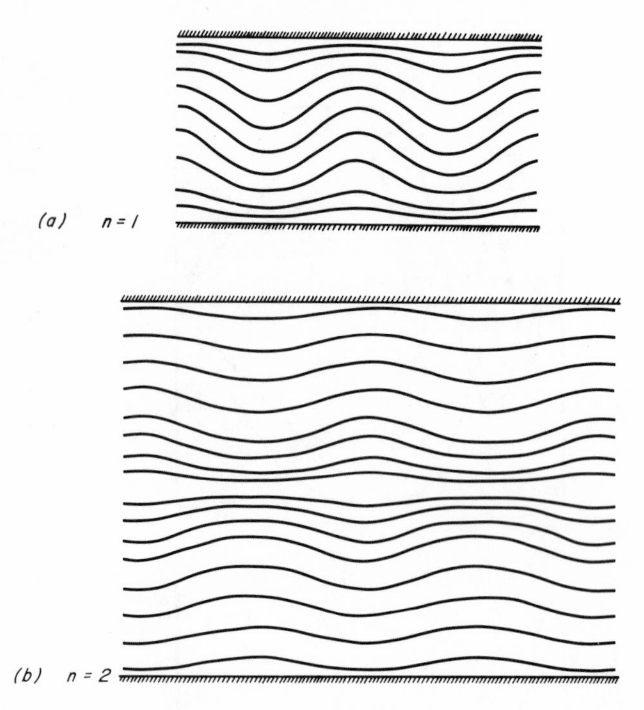

FIG. 3-4. Wave profiles for internal waves of the first five modes showing the isopycnals. All the waves are of finite amplitude, i.e., the assumption $ak \ll 1$ does not hold. (a) A first-order internal wave in a fluid with a linear density profile (redrawn from Thorpe, 1968c); (b) a second-order internal wave in a fluid with a tanh density profile (adapted from Thorpe, 1968c); (c), (d), (e) internal waves of order three through five in a fluid with a linear density profile. The isopycnals are sketched from photographs of wavetank experiments (Orlanski, 1972).

$$\nu_n = \frac{(k_h)_n^2}{\omega^2} = \frac{1}{(c_p)_n^2} \tag{9}$$

where $\vec{c}_p \equiv \frac{\omega}{k_h^2}\vec{k}_h$ is the phase velocity for a wave of frequency ω and wavenumber $\vec{k}_h$. Yih (1969) has given a summary of theorems concerning $\vec{c}_p$ for two-dimensional waves.

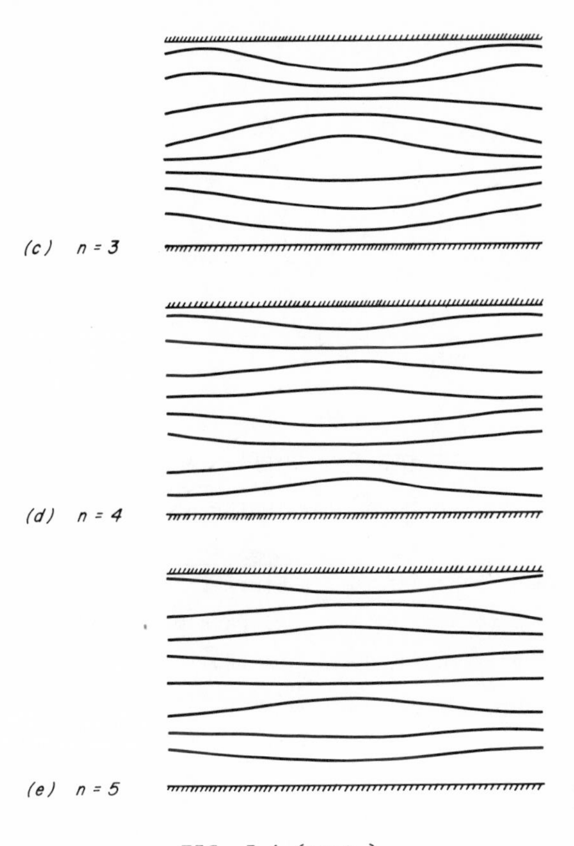

FIG. 3-4 (cont.).

If we assume that there is no y-dependence in the motion, so that $v \equiv 0$ and $F(x,y) = F(x)$, and if we assume

$$w = F(x)\mathcal{W}(z)\exp(-i\omega t)$$

$$u_x = G(x)\mathcal{U}(z)\exp(-i\omega t)$$

then from the equation of continuity, $\partial u_x/\partial x = -\partial w/\partial z$, we have

$$\mathcal{U}(z)\int_0^x \frac{dG}{dx}\,dx = -\frac{d\mathcal{W}}{dz}\int_0^x F(x)\,dx$$

Thus, at a specific point x_o,

$$\mathcal{U}(z)G(x)\Big|_0^{x_o} = -\frac{d\mathcal{W}}{dz}F(x)\Big|_0^{x_o}$$

so that

$$\mathcal{U}(z) = -\frac{F(x_o) - F(0)}{G(x_o) - G(0)}\frac{d\mathcal{W}}{dz}$$

or

$$\mathcal{U}(z) \quad d\mathcal{W}$$

where $C_1 = -\dfrac{F(x_o) - F(0)}{G(x_o) - G(0)}$ = constant. Thus, the first derivative of $\mathcal{W}(z)$ is directly proportional to the horizontal velocity (Krauss, 1966b, p. 31).

Equations (8) constitute a Sturm-Liouville equation, and they are often used to discuss internal wave motion (Fjeldstad, 1933; Yih, 1960, 1969; Krauss, 1966b, p. 30; Thorpe, 1968c). If $\left|\dfrac{d^2\mathcal{W}}{dz^2}\right| >> \left|\dfrac{N^2}{g}\dfrac{d\mathcal{W}}{dz}\right|$ (Krauss, 1966b, p. 34), then (8a) becomes

$$\frac{d^2\mathcal{W}}{dz^2} + \frac{N^2 - \omega^2}{\omega^2 - f^2}k_h^2\mathcal{W} = 0 \tag{11}$$

If, in addition, we assume

$$\omega^2 >> f^2 \tag{12}$$

i.e., if the period of the waves is much less than the inertial period, then Eq. (8a) becomes

$$\frac{d^2\mathcal{W}}{dz^2} + \frac{N^2 - \omega^2}{\omega^2}k_h^2\mathcal{W} = 0 \tag{13}$$

which has been discussed by Phillips (1966, p. 162) and Eckart (1961, p. 793).

Equations (8) are very basic. Many people have solved them for different distributions of N(z), and we proceed now to discuss these solutions.

Homogeneous ocean

If the density is constant from the surface to the bottom of the ocean, then $d\bar{\rho}/dz = 0$, so than $N \equiv 0$. If $\omega^2 > f^2$, then

$$\frac{\omega^2 - f^2}{-\omega^2} < 0$$

and internal gravity waves are not possible (Sec. 3.3). We are excluding from consideration the case $\omega \leq f$. Inertial waves, however, are possible in a homogeneous, rotating fluid (Veronis, 1967; Phillips, 1966, pp. 192-193).

N constant: plane waves (Fig. 3-5)

Because N = constant makes Eqs. (8) much more tractable, and because the exponential profile (or linear profile, depending on how N is defined, see Sec. 3.3) is an excellent approximation to the density profile of the atmosphere, this case is often used in mathematical analyses of internal waves. The validity of the approximation to the actual density of the ocean is somewhat questionable. For the shallow Baltic Sea [where most internal waves are describable using only the first three modes (Krauss, 1972)], the actual density profile is very nearly exponential (Krauss, 1966b, p. 34). Likewise, N = constant may, under certain circumstances, be an acceptable local approximation to a diffuse thermocline (Phillips, 1966, p. 174). Internal waves in deep water, however, may need five or ten eigenfunctions to describe them (Krauss, 1972), and Cox and Sandstrom (1972) have shown that the eigenfunctions in a fluid with an exponential density profile bear no resemblance to those of the real ocean after the first few modes.

If N = constant, Eq. (8a) becomes

$$\frac{d^2\mathcal{W}}{dz^2} - \frac{N^2}{g}\frac{d\mathcal{W}}{dz} + \left(\frac{N^2 - \omega^2}{\omega^2 - f^2}\right) k_h^2 \mathcal{W} = 0 \qquad (14)$$

We allow the waves to propagate in three dimensions and assume that the vertical velocity w has the form

$$w(x,y,z,t) = \mathcal{W}(z)\exp[i(\vec{k}_h \cdot \vec{r} - \omega t)] \qquad (15)$$

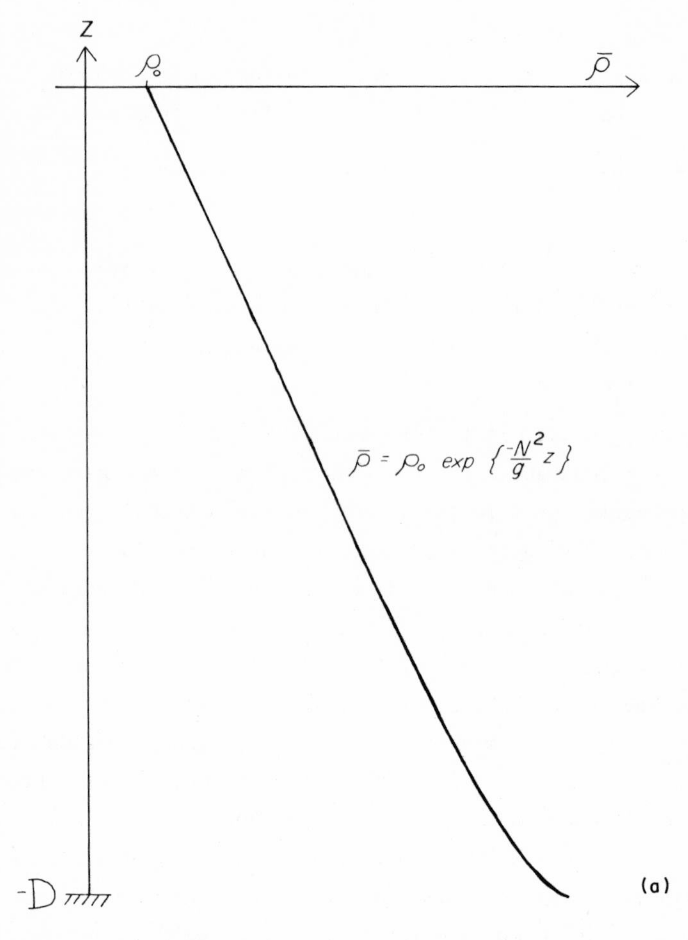

FIG. 3-5. (a) Density profile (from Table 3-1) and (b) Brunt-Väisälä frequency for the case N = constant.

where

$$\mathcal{W}(z) = C_o \exp[(N^2/2g)z]\left\{ B_1 \exp(ik_z z) + B_2 \exp(-ik_z z)\right\} \quad (16)$$

(Phillips, 1966, p. 174; Yih, 1969, p. 86.)
Then $\mathcal{W}(z)$ given by Eq. (16) is a solution of Eq. (14), if we require that

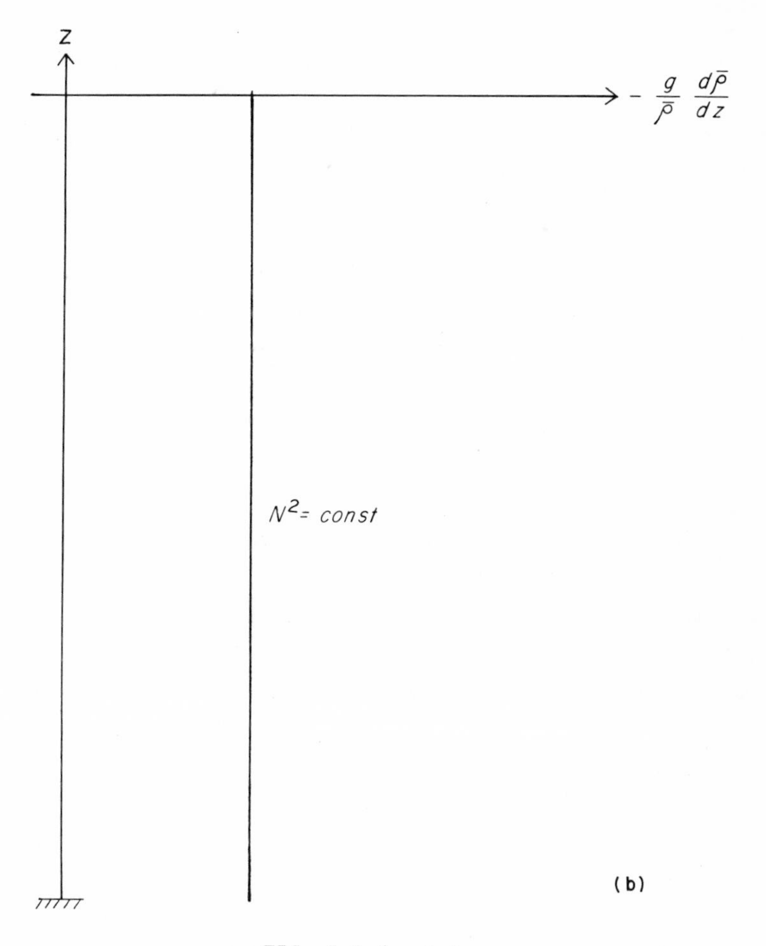

FIG. 3-5 (cont.).

$$k_z^2 = \left(\frac{N^2 - \omega^2}{\omega^2 - f^2}\right) k_h^2 - \left(\frac{N^2}{2g}\right)^2 \tag{17}$$

This may be shown by direct calculation. We now apply the bottom and surface boundary conditions, Eqs. (8b) and (8c), respectively. [Lamb (1945, p. 379) has discussed the solution for a free surface.] The surface boundary condition gives

$$B_1 = -B_2 \tag{18}$$

while Eq. (8b) gives

$$\exp(2ik_z D) = 1$$

so that

$$k_z = \frac{n\pi}{D} \tag{19}$$

Hence, letting $C_1 = C_o B_1$, Eq. (16) becomes

$$\begin{aligned} \mathcal{W}_n(z) &= C_1 \, e^{(N^2/2g)z}[e^{(n\pi iz/D)} - e^{-(n\pi iz/D)}] \\ &= 2iC_1 e^{(N^2/2g)z} \sin(n\pi z/D) \end{aligned} \tag{20}$$

(Krauss, 1966b, p. 33; Kenyon, 1968; Martin, Simmons and Wunsch, 1972). The eigenfunctions for the first three modes are shown in Fig. 3-6.

Equation (17) gives a dispersion relation which is independent of the boundary conditions. If $f^2 << \omega^2$, that is, if the waves are of short period, so that the earth's rotation may be neglected, Eq. (17) becomes

$$\left(\frac{\omega}{N}\right)^2 = \frac{k_h^2}{k^2 + (N^2/2g)^2} \tag{21}$$

since $k^2 = k_h^2 + k_z^2$. Short-period waves also have short wavelengths, i.e. large wavenumbers, so that the approximation $(N^2/2g)^2 << k^2$ holds in this case. Then Eq. (21) becomes

$$\begin{aligned} \left(\frac{\omega}{N}\right)^2 &= \frac{k_h^2}{k^2} \\ &= \frac{k_h^2}{k_h^2 + k_z^2} \end{aligned} \tag{22}$$

so that

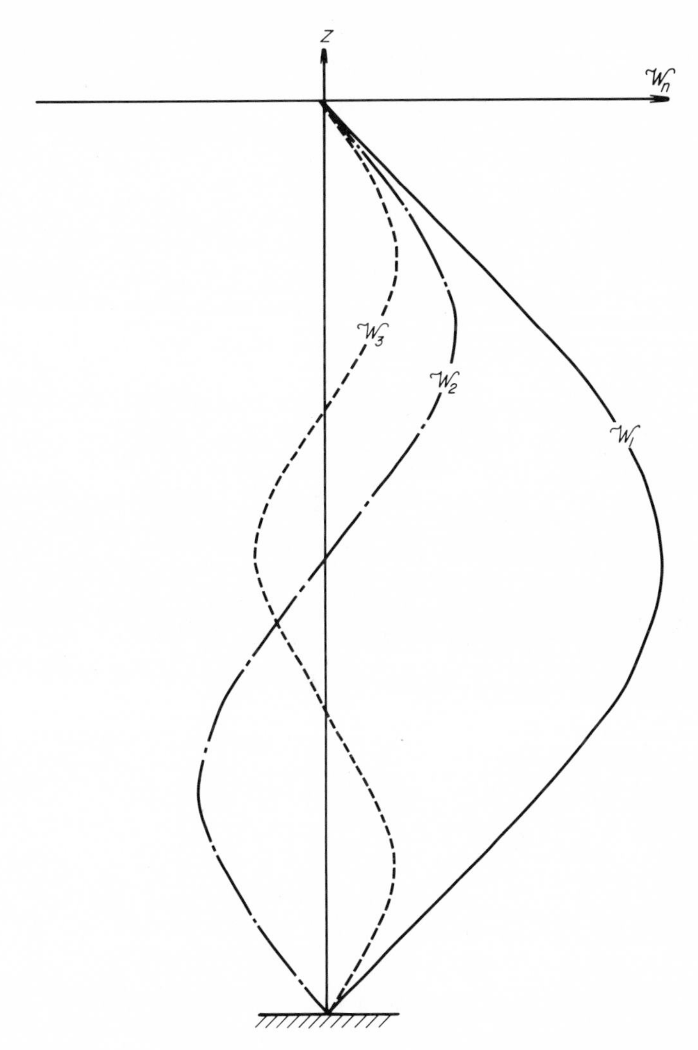

FIG. 3-6. Eigenfunctions for the first three internal wave modes when N is constant.

$$\left(\frac{\omega}{N}\right)^2 = \frac{k_h^2}{k_h^2 + (n\pi/D)^2} \qquad \text{Dispersion relation for short waves in water of depth -D.} \qquad (23)$$

As may be seen from Fig. 3-7.,

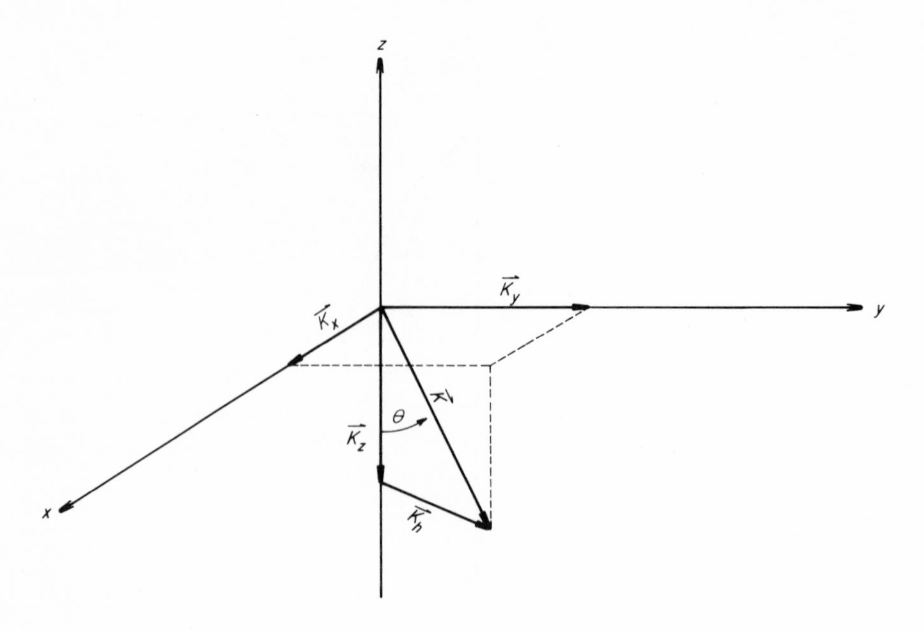

FIG. 3-7. Vector-angle relations for the wavenumber vector, $\vec{k}$.

$$k_h = k \sin \theta \qquad (24)$$

so that Eq. (22) may also be written

$$\omega = \pm N \sin \theta \qquad (25)$$

[Phillips (1966, p. 174) takes θ to be the angle between the wavenumber and the horizontal, so that $\omega = \pm N \cos \theta$.]

Equation (25) may be obtained more directly (Hughes 1972: personal communication; Bretherton, 1971). In Fig. 3-8, consider a particle displaced an arbitrary amount s along a wavefront $\vec{c}_g$ so that

$$\zeta = s \sin \theta \qquad (26)$$

From fluid mechanics, the magnitude of the buoyancy force $\vec{F}_b$ acting upon the particle is

$$\left|\vec{F}_b\right| = \left|g\zeta \frac{d\bar{\rho}}{dz}\right|$$

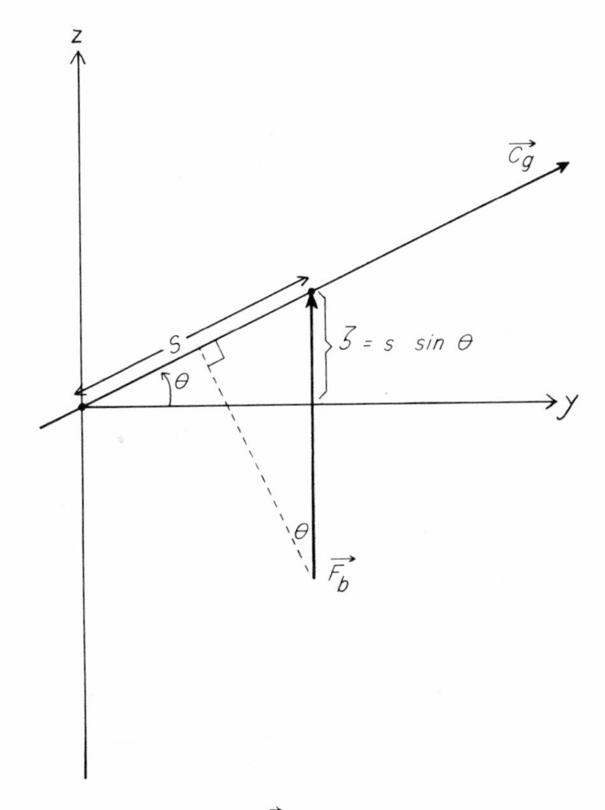

FIG. 3-8. Buoyancy force $\vec{F}_b$ acting on a particle displaced by an amount s along a wavefront $\vec{c}_g$.

Thus, the magnitude of the force acting on the particle per unit mass is

$$\left|\frac{\vec{F}_b}{\bar{\rho}}\right| = \left|-N^2\zeta\right|$$

and the component of $\vec{F}_b/\bar{\rho}$ along the wavefront $\vec{c}_g$ is

$$\begin{aligned}(F_b/\bar{\rho}) \sin\theta &= -N^2 \zeta \sin\theta \\ &= -(N^2 \sin^2\theta)\, s \qquad (27)\end{aligned}$$

Along the wavefront there is no variation in pressure, so that Eq. (27) must be the total force on the particle along the wavefront, and from $F = ma$, we have

$$\frac{d^2 s}{dt^2} = -(N^2 \sin^2\theta)\, s$$

or

$$\frac{d^2 s}{dt^2} + (N^2 \sin^2\theta)\, s = 0$$

In analogy with a simple harmonic oscillator, the frequency ω of the particle's oscillation is given by $\omega = \pm N \sin\theta$.

In the following, we assume that the earth's rotation may be neglected. Phillips (1966, pp. 193-195) has discussed the case when the earth's rotation is included. The *phase velocity* $\vec{c}_p$ is defined

$$\vec{c}_p \equiv \frac{\omega}{k^2}\vec{k} = \frac{\omega}{k^2}[\hat{i}k_x + \hat{j}k_y + \hat{k}k_z] = \hat{i}c_{p_x} + \hat{j}c_{p_y} + \hat{k}c_{p_z} \tag{28}$$

[Note that this definition of $\vec{c}_p$ is not the same as the definition of c_p as ω/k_x, which has been found useful in the reduction of three-dimensional normal modes to two-dimensional ones (Yih, 1969, p. 86; Bretherton, 1970, pp. 62-63).]

From Eq. (21),

$$c_p^2 = \frac{\omega^2}{k^2} = \frac{N^2 k_h^2}{k^2[k^2 + (N^2/2g)^2]} \tag{29}$$

The *group velocity* $\vec{c}_g$ is defined

$$\vec{c}_g \equiv \hat{i}\frac{\partial\omega}{\partial k_x} + \hat{j}\frac{\partial\omega}{\partial k_y} + \hat{k}\frac{\partial\omega}{\partial k_z} = \hat{i}c_{g_x} + \hat{j}c_{g_y} + \hat{k}c_{g_z} \tag{30}$$

Using Eq. (21), we compute the components of $\vec{c}_g$

$$c_{g_x} = \pm N\left\{\frac{k_z^2 + (N^2/2g)^2}{k_h[k^2 + (N^2/2g)^2]^{3/2}}\right\} k_x \tag{31a}$$

$$c_{g_y} = \pm N \left\{ \frac{k_z^2 + (N^2/2g)^2}{k_h[k^2 + (N^2/2g)^2]^{3/2}} \right\} k_y \tag{31b}$$

$$c_{g_z} = \pm N \left\{ \frac{-k_h}{[k^2 + (N^2/2g)^2]^{3/2}} \right\} k_z \tag{31c}$$

Since $c_{g_y}/c_{g_x} = k_y/k_x = c_{p_y}/c_{p_x}$, $\vec{c}_g$ and $\vec{c}_p$ are in the same vertical plane (Yih, 1969, p. 87).

For short waves $N^2/2g \ll k$ and in this case

$$c_{g_x} = \pm N \left(\frac{k_z^2}{k_h k^3} \right) k_x$$

$$c_{g_y} = \pm N \left(\frac{k_z^2}{k_h k^3} \right) k_y \tag{32}$$

$$c_{g_z} = \pm N \left(\frac{-k_h}{k^3} \right) k_z$$

It may be directly computed that the phase velocity $\vec{c}_p$ as given by Eq. (28) is perpendicular to the group velocity $\vec{c}_g$ as given by Eqs. (32).

$$\vec{c}_p \cdot \vec{c}_g = \frac{\pm\omega N}{k^2} \left[\frac{k_x^2 k_z^2}{k_h k^3} + \frac{k_y^2 k_z^2}{k_h k^3} - \frac{k_h k_z^2}{k^3} \right]$$

$$= \frac{\pm\omega N}{k^2} \left[\frac{(k_x^2 + k_y^2) k_z^2 - k_h^2 k_z^2}{k_h k^3} \right]$$

$$= 0$$

and thus $\vec{c}_p$ is perpendicular to $\vec{c}_g$. This may be difficult to visualize at first. The wavetank experiments of Mowbray and Rarity (1967a), Sec. 1.7, are helpful in seeing what happens (Figs. 3-7 and 3-8 show the same angle θ as Fig. 1-6).

Sharp thermocline (Fig. 3-9)

This case frequently occurs in tropical or subtropical waters when the upper layer of the ocean has been well-mixed by the wind. With this particular model, no analytic expression is specified for the thermocline region. Care must be used in making the Boussinesq approximation when this model is used, particularly when a finite depth is assumed (Benjamin, 1967). It seems that the Boussinesq

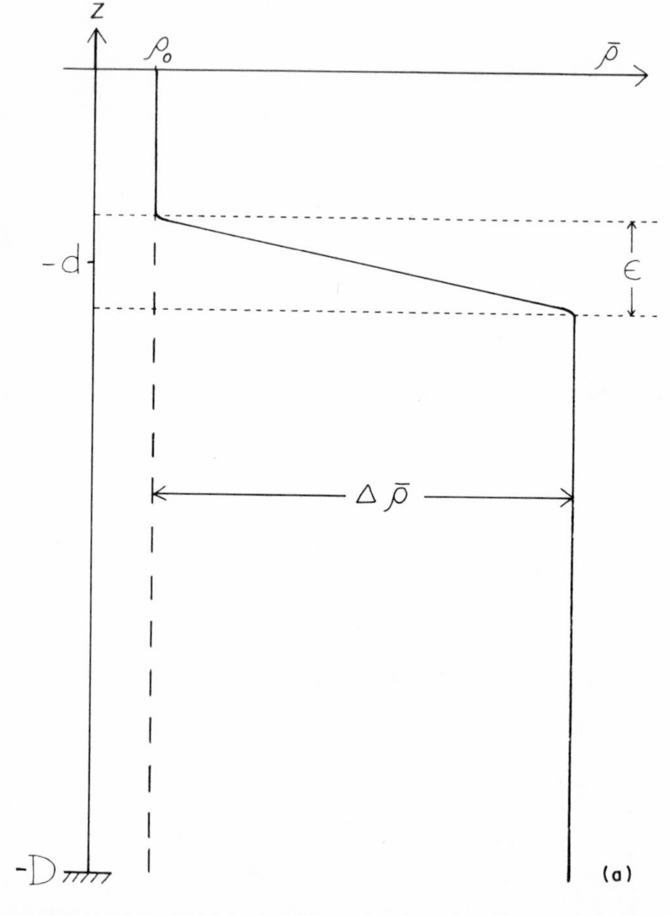

FIG. 3-9. (a) Density and (b) Brunt-Väisälä frequency for a fairly sharp thermocline separating nearly homogeneous water masses above and below.

approximation is not valid for waves which are sufficiently long compared to the thermocline thickness ε (Thorpe, 1968c, p. 566). Phillips (1966, pp. 166-173) has given a detailed discussion of the propagation of a first-order internal wave in such a system. Taking $\left|\frac{d^2 \mathcal{W}}{dz^2}\right| >> \left|\frac{N^2}{g}\frac{d\mathcal{W}}{dz}\right|$ and assuming that the waves are short enough so that the effect of the earth's rotation may be neglected, the equation to be solved is Eq. (13), with N(z) = 0. The first eigenfunction is given by

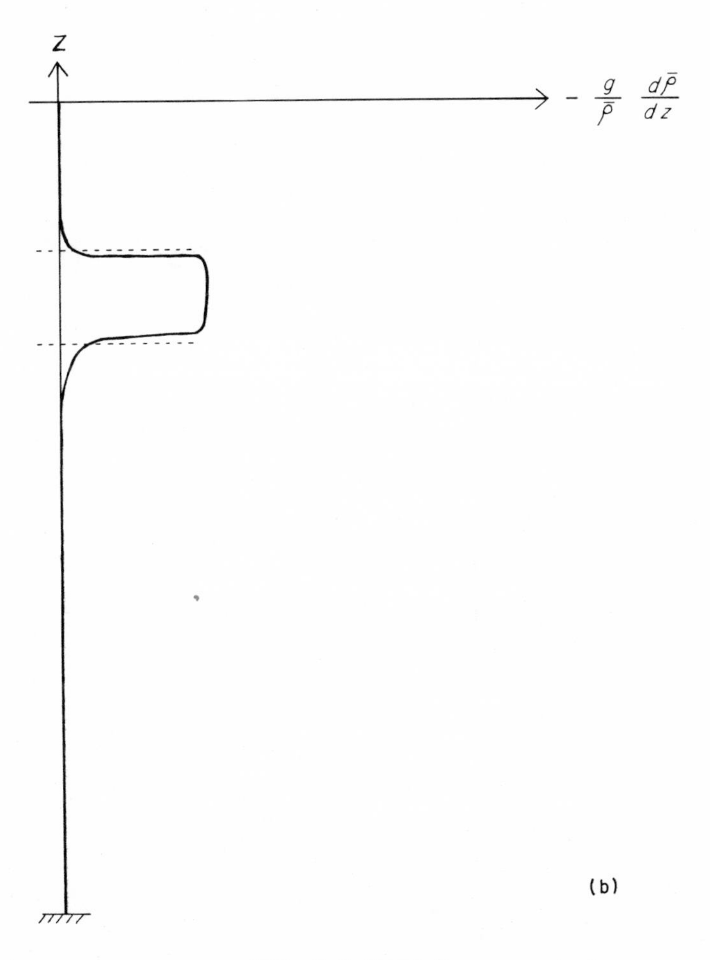

FIG. 3-9 (cont.).

$$\mathcal{W}(z) = \begin{cases} C_1 \sinh k_h^2 z & 0 \geqslant z > -d \\ \\ C_2 \sinh k_h(z + D) & -d > z \geqslant -D \end{cases} \tag{33}$$

(Fig. 3-10).

The dispersion relation is given by

$$\omega^2 - gk_h \frac{\Delta\bar{\rho}}{\rho_0} \left(k_h \varepsilon + \coth k_h d + \coth k_h (D-d) \right)^{-1} \doteq 0$$

(Phillips, 1966, p. 167. If the length of the internal waves is much

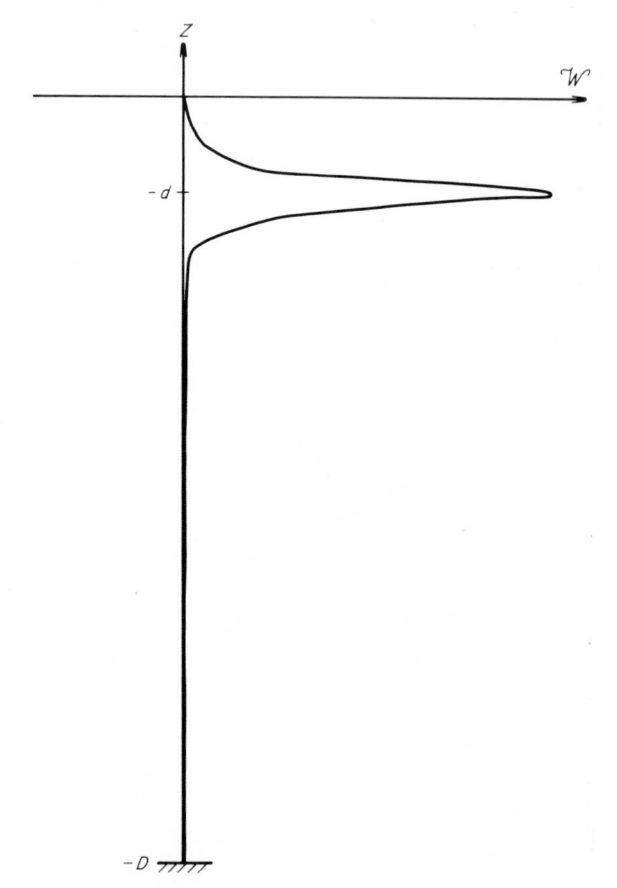

FIG. 3-10. The first eigenfunction in the case of a sharp thermocline.

greater than the thermocline thickness but much less than the distance (D-d), this reduces to

$$\omega^2 - gk_h \frac{\Delta\bar{\rho}}{\rho_0} \left(1 + \coth k_h d \right)^{-1} = 0 \tag{34}$$

Phillips has given expressions for the vertical displacement of the thermocline and of the surface, for the particle velocities u_x and w in the thermocline and for u_x at the surface, for the wave shear across the thermocline, and for the internal wave energy and wave momentum.

Benjamin (1967), in examining the properties of solitary waves and periodic waves of permanent form, has treated the similar case of a sharp thermocline between two infinite, homogeneous water masses.

Tanh density profile (Fig. 3-11)

The tanh density profile

$$\bar{\rho} = \rho_{-d} \exp\left[\frac{\Delta\bar{\rho}}{2g\rho_{-d}} \tanh 2\varepsilon^{-1}(-z-d) \right]$$

was first described by Groen (1948a), and is a fairly good model of the actual density profile of the ocean. As with the previous model, the Boussinesq approximation acts to restrict the length of waves for which the theory is valid (Thorpe, 1968c).

Taking $\left| \frac{d^2 \mathcal{W}}{dz^2} \right| >> \left| \frac{N^2}{g} \frac{d\mathcal{W}}{dz} \right|$, an expression for the eigenfunctions of Eq. (11) may be found in terms of hypergeometric functions (Groen, 1948a; Krauss, 1966b, p. 36). The first three eigenfunctions are shown in Fig. 3-12.

The dispersion relation for this model is given by

$$k_h^2 + \frac{2\varepsilon^{-1}(2n+1)(\omega^2-f^2)^{1/2}\,\omega}{\omega^2 - g\varepsilon^{-1}(\Delta\bar{\rho}/\rho_{-d})} k_h + \frac{\varepsilon^{-2}\,[(2n+1)^2-1]\,(\omega^2-f^2)}{\omega^2 - g\varepsilon^{-1}(\Delta\bar{\rho}/\rho_{-d})} = 0 \tag{35}$$

Krauss, 1966b, p. 37).

Krauss (1966b, pp. 35-38) and Thorpe (1968c), as well as Groen, have given a discussion of this model; Miles (1972) and Davis and Acrivos (1967b) have treated it briefly.

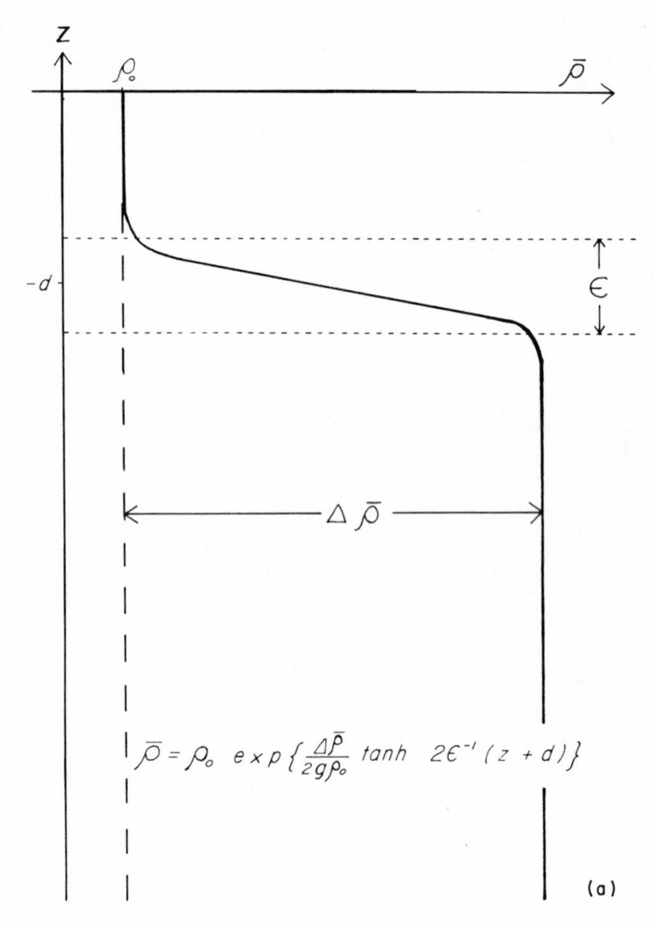

FIG. 3-11. (a) Density and (b) Brunt-Väisälä frequency for the case of a tanh density profile.

Exponential-thermocline profile (Fig. 3-13)

This model has been considered by Fjeldstad (1933), Krauss (1966b, pp. 41-43), and Keller and Munk (1970). If the rotation of the earth is neglected and if $\left|\frac{d^2\mathcal{W}}{dz^2}\right| \gg \left|\frac{N^2}{g}\frac{d\mathcal{W}}{dz}\right|$, then the equation for which we seek eigensolutions is Eq. (13). These are given by

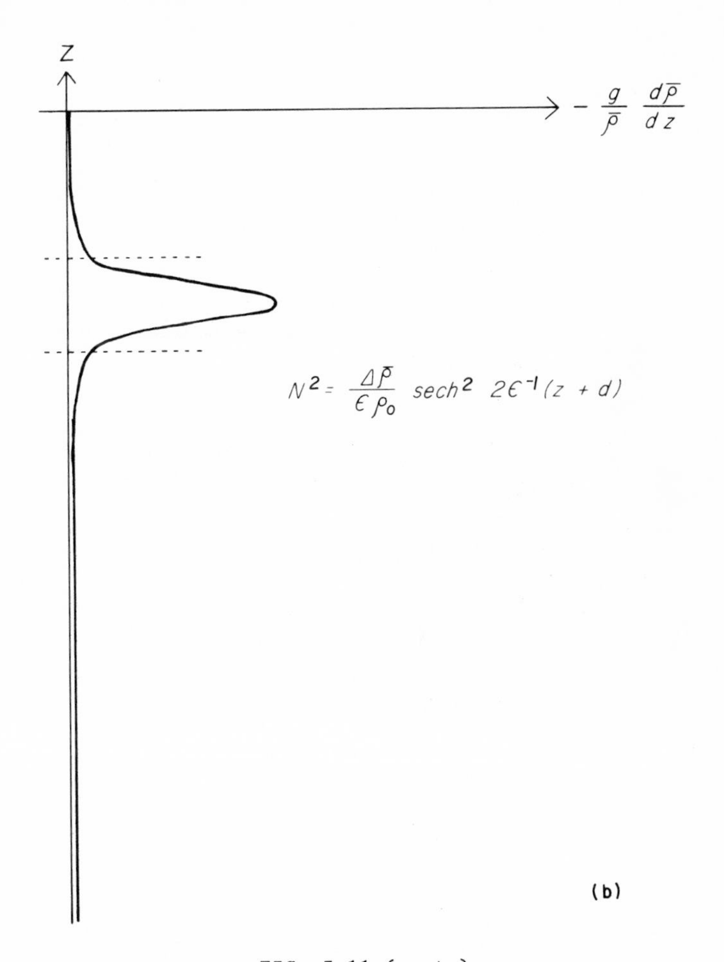

FIG. 3-11 (cont.).

$$\mathcal{W}_n(z) = \begin{cases} C_1 \sinh k_{h_n} z & 0 \geqslant z \geqslant z_1 \\ C_2 \sin\left[k_{h_n}\left(\dfrac{N^2-\omega^2}{\omega^2}\right)^{1/2} z\right] + C_3 \cos\left[k_{h_n}\left(\dfrac{N^2-\omega^2}{\omega^2}\right)^{1/2} z\right] & \\ & z_1 \geqslant z \geqslant z_2 \\ C_4 \exp(k_{h_n} z) + C_5 \exp(-k_{h_n} z) & z_2 \geqslant z \geqslant -D \end{cases} \tag{36}$$

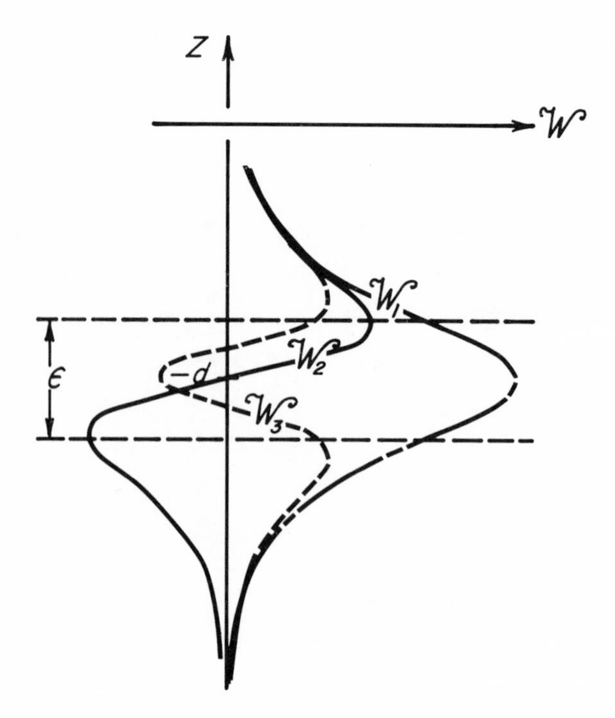

FIG. 3-12. The first three eigenfunctions for a tanh density model. (Redrawn from Groen, 1948a.)

Krauss (1966b, p. 42) has set up the equations for obtaining the constants C_1, C_2, C_3, C_4, and C_5. Fjeldstad (1933) has graphed the eigenfunctions $\mathcal{W}_n(z)$ and the horizontal particles speeds $\mathcal{U}_n(z)$ as a function of depth for n=1,2,3 (Fig. 3-14). Fjeldstad (1933) has used Eq. (10) for computing $\mathcal{U}(z)$.

The dispersion relation is given by

$$\tan\left[k_h\left(\frac{N^2-\omega^2}{\omega^2}\right)^{1/2}\right]\varepsilon + \frac{\omega\left(\frac{N^2-\omega^2}{\omega^2}\right)^{1/2}[\tanh k_h z_1 - \tanh k_h(z_2+D)]}{1+\left(\frac{N^2-\omega^2}{\omega^2}\right)\tanh k_h z_1 \tanh k_h(z_2+D)} = 0 \tag{37}$$

[Krauss, 1966b, p. 42; Keller and Munk, 1970, their Eq. (48) with $\alpha \equiv 0$]. For the deep ocean and for higher mode internal waves, the approximation $k_h D >> 1$ can be expected to hold, and Eq. (37) becomes

$$\omega^2 - gk_h \frac{\Delta\bar{\rho}}{\rho_0} \left(\frac{1}{n\pi} - \frac{\varepsilon^{k_h}}{n^3 \pi^3} + \cdots \right) = 0 \qquad n >> 1 \tag{38}$$

(Keller and Munk, 1970).

Thorpe (1968c, pp. 591-592) has given a brief discussion of the dispersion relation for a system with an exponential thermocline between two infinite homogeneous fluids.

Three-fluid system with $N = N_1, N_2, N_3$ (Fig. 3-15)

Kanari (1968) has used this model in discussing internal waves in Lake Biwa, Japan. Taking $\left|\frac{d^2 \mathcal{W}}{dz^2}\right| >> \left|\frac{N^2}{g} \frac{d\mathcal{W}}{dz}\right|$ and neglecting the earth's rotation, the equation to be solved is Eq. (13); its eigensolutions are given by

$$\mathcal{W}_n(z) = \begin{cases} C_1 \dfrac{\sin[k_{h_n}(N_1/\omega) z]}{\sin[k_{h_n}(N_1/\omega) z_1]} & 0 \geqslant z \geqslant z_1 \\[2ex] \dfrac{C_1 \sin[k_{h_n}(N_2/\omega)(z_2 - z)] + C_2 \sin[k_{h_n}(N_2/\omega)(z - z_1)]}{\sin[k_{h_n}(N_2/\omega)(z_2 - z_1)]} & z_1 \geqslant z \geqslant z_2 \\[2ex] C_2 \dfrac{\sin[k_{h_n}(N_3/\omega)(-D - z)]}{\sin[k_{h_n}(N_3/\omega)(-D - z_2)]} & z_2 \geqslant z \geqslant -D \end{cases} \tag{39}$$

where the constants have been chosen so that the eigensolutions $\mathcal{W}_n(z)$ agree at the boundaries z_1 and z_2. The eigensolutions for $n = 1,2,3,4$ are shown in Fig. 3-16.

The dispersion relation is given by

$$N_2^2 \cdot \frac{\tan(\alpha k_{h_n}/\omega)}{N_1} \cdot \frac{\tan(\beta k_{h_n}/\omega)}{N_2} \cdot \frac{\tan(\sigma k_{h_n}/\omega)}{N_3}$$

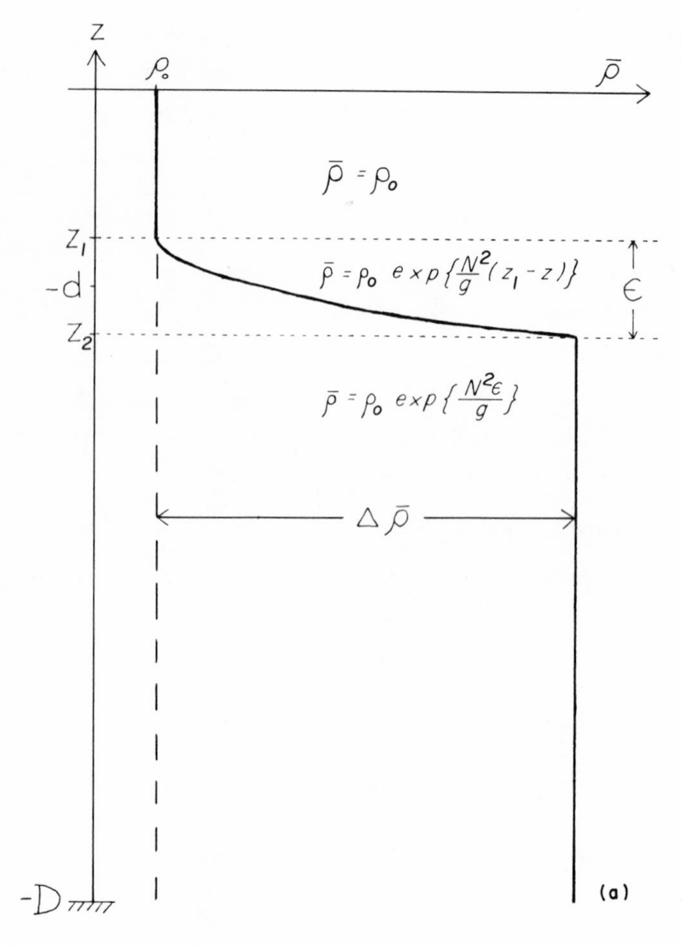

FIG. 3-13. (a) Density and (b) Brunt-Väisälä frequency for the case of an exponential-thermocline profile.

$$-\left[\frac{\tan(\alpha k_{h_n}/\omega)}{N_1} + \frac{\tan(\beta k_{h_n}/\omega)}{N_2} + \frac{\tan(\sigma k_{h_n}/\omega)}{N_3}\right] = 0 \tag{40}$$

where

$$\alpha = N_1 z_1 \qquad \beta = N_2(z_2 - z_1) \qquad \sigma = N_3(-D - z_2)$$

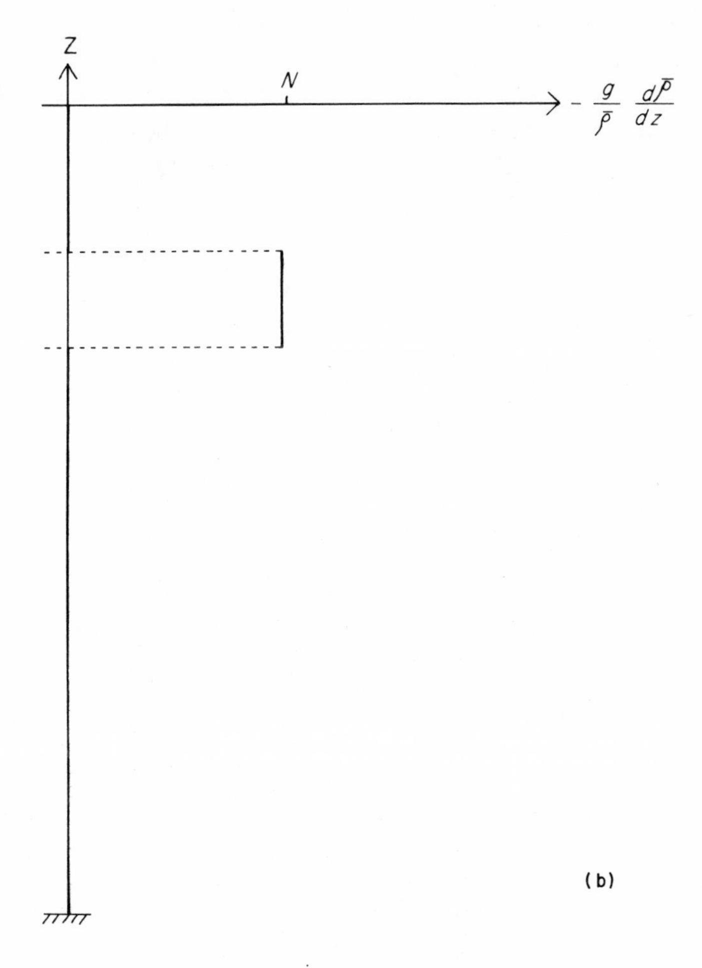

FIG. 3-13 (cont.).

N exponential (Fig. 3-17)

This model has been considered by Tareev (1963), and more recently by Garrett and Munk (1972b). It is particularly typical of tropical waters, where the density is nearly constant above 100 m and below 3000 m. Yearsley (1966) has used a somewhat similar density distribution given by $\bar{\rho} = \rho_0(1-\alpha e^{\beta z})$ for z from 0 to -D. Below the mixed

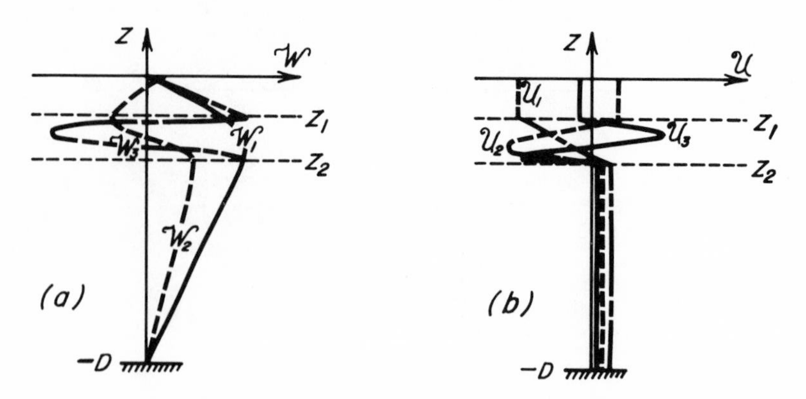

FIG. 3-14. (a) Eigenfunctions and (b) horizontal particle speeds for the case of an exponential thermocline between two homogeneous layers. (Redrawn from Fjeldstad, 1933.)

layer, Yearsley's distribution agrees fairly well with density observations from the Arctic Ocean.

The approximation for N^2 is derived as follows:

$$-\frac{g}{\bar{\rho}}\frac{d\bar{\rho}}{dz} = -\frac{g}{\bar{\rho}}(-\Delta\bar{\rho}\ 2\beta e^{2\beta d}e^{2\beta z}) = \frac{g\Delta\bar{\rho}\ 2\beta e^{2\beta d}e^{2\beta z}}{\rho_0+\Delta\bar{\rho}(1-e^{2\beta d}e^{2\beta z})}$$

$$\doteq \frac{g\Delta\bar{\rho}\ 2\beta e^{2\beta d}e^{2\beta z}}{\rho_0} = \alpha^2 e^{2\beta z}$$

by the definition of $\Delta\bar{\rho}$. Garrett and Munk have avoided the problem by using $N^2 \equiv -\frac{g}{\rho_0}\frac{d\bar{\rho}}{dz}$ (Sec. 3.3).

As usual, we take $\left|\frac{d^2\mathcal{W}}{dz^2}\right| >> \left|\frac{N^2}{g}\frac{d\mathcal{W}}{dz}\right|$, so that the equation of interest is Eq. (11). If we let $\mathcal{W}_n(z) = \eta_n(r)$, where $r = \alpha\sqrt{\nu_h}e^{\beta z}/\beta$, and let $m = \omega\sqrt{\nu_h}/\beta$, then Eq. (11) becomes

$$r^2\frac{d^2\eta_n}{dr^2} + r\frac{d\eta_n}{dr} + (r^2-m^2)\eta_n = 0$$

which is Bessel's equation, with solution

$$\eta_n(r) = C_3 J_m(r) + C_4 Y_m(r)$$

Thus, the eigenfunctions of Eq. (11) with boundary conditions (8b) and (8c) are given by

$$\mathcal{W}_n(z) = \begin{cases} C_1 \sinh(\omega\sqrt{\nu_n}\, z) & 0 \geqslant z \geqslant -d \\ C_2 \left[J_m(r) - \dfrac{J_m(r_{-D})}{Y_m(r_{-D})} Y_m(r) \right] \equiv C_2 \mathcal{J}_m(r) & -d \geqslant z \geqslant -D \end{cases} \tag{41}$$

Here J and Y are Bessel functions of the first and second kind, respectively, ν_n is the nth eigenvalue, given by Eq. (7), C_1 and C_2 are arbitrary constants, and $r_{-D} = \alpha\sqrt{\nu_n}\, e^{-\beta D}/\beta$. The eigensolutions for modes 1 and 5 are shown in Fig. 3-18 for the cases $\omega/N = 0.75$ and $\omega/N = 0.03$.

Continuity of $\mathcal{W}_n$ and of $d\mathcal{W}_n/dz$ at $z = -d$ requires that

$$-C_1 \sinh(\omega\sqrt{\nu_n}\, d) = C_2 \mathcal{J}_m(r_{-d}) \tag{42}$$

and

$$C_1\, \omega\sqrt{\nu_n} \cosh(\omega\sqrt{\nu_n}\, d) = C_2 \frac{d}{dz}\left[\mathcal{J}_m(r_{-d})\right] \tag{43}$$

Combining Eqs. (42) and (43), we obtain the dispersion relation

$$\omega\sqrt{\nu_n} \coth(\omega\sqrt{\nu_n}\, d) + \frac{1}{\mathcal{J}_m(r_{-d})} \frac{d}{dz}\left[\mathcal{J}_m(r_{-d})\right] = 0 \tag{44}$$

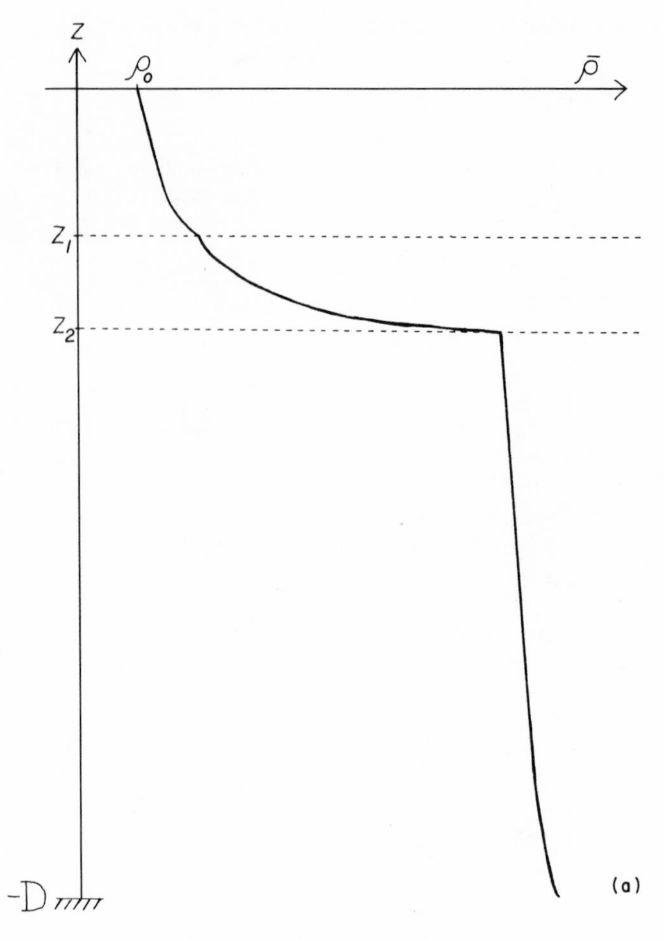

FIG. 3-15. (a) Density and (b) Brunt-Väisälä frequency for the case of three superposed, exponentially stratified layers.

Special exponential density distribution (Fig. 3-19)

Internal wave solutions for this model have been discussed by Krauss (1966b, pp. 38-39). Neglecting $(N^2/g)(d\mathcal{W}/dz)$, the equation to be solved is Eq. (11). The eigensolutions may be given in terms of a Whittaker function, $M_{k,\sigma/2}$

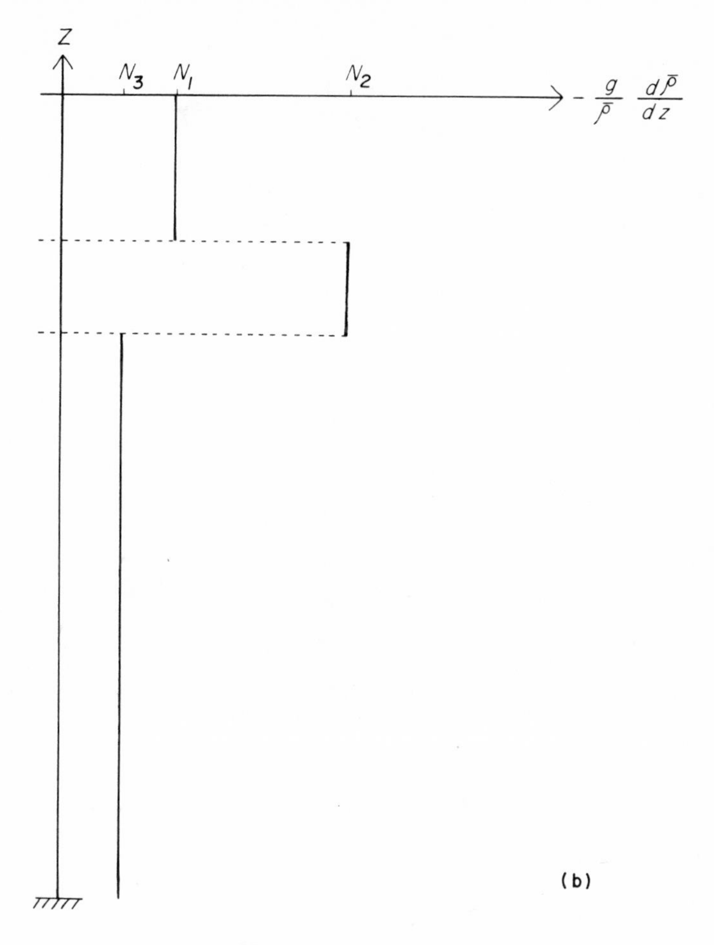

FIG. 3-15 (cont.).

$$\mathscr{W}_n(z) = C_1\ z^{-1/2}\ M_{K_n,+\frac{1}{4}}\left[(3\beta g\nu_n)^{1/2}\ z^2\right] \tag{45}$$

where

$$K_n = \frac{g\alpha-\omega^2}{4}\left(\frac{\nu_n}{3\beta g}\right)^{1/2}$$

The Whittaker function

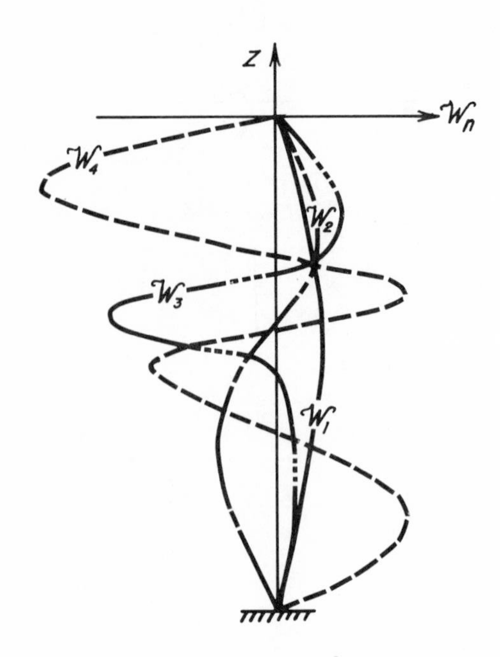

FIG. 3-16. Eigensolutions (n = 1,2,3,4) for a three-layered system with N constant within each layer. (Redrawn from Kanari, 1968.)

$$M_{K,\sigma/2}(x) \equiv \frac{x^{(1+\sigma)/2}\ e^{-x/2}\ F\{[(1+\sigma)/2] - K;\ 1+\sigma;\ x\}}{\Gamma(1+\sigma)}$$

where $\Gamma(1+\sigma)$ is the gamma function, and F is a hypergeometric function

$$F(A,B,x) \equiv 1 + \frac{A}{B}x + \frac{A(A+1)x^2}{B(B+1)2!} + \frac{A(A+1)(A+2)x^3}{B(B+1)(B+2)3!} + \cdots$$

Figure 3-20 shows the first four eigenfunctions given by Eq. (45).

The dispersion relation in this system is given by

$$M_{K,+\frac{1}{4}}[(3\beta g\nu_n)^{1/2}\,D^2] = 0 \tag{46}$$

The eigenvalues ν_n, and hence the dispersion relation, may be obtained from Eq. (46) by approximation.

Quasilinear density profile (Fig. 3-21)

This model has been used by Magaard (1962) to describe internal waves propagating over an uneven bottom topography, in particular, to show what happens when they propagate into shallow water. Krauss (1966b, p. 41) has summarized some of Magaard's results.

The eigensolutions to Eq. (8) for long waves ($\omega^2 << N^2$) are given by

$$\mathcal{W}_n(z) = a_n \sin\left[n\pi \frac{(C_1-2z)^{1/2} - (C_1)^{1/2}}{(C_1+2D)^{1/2} - (C_1)^{1/2}} \right] \tag{47}$$

if

$$\frac{gk_h^2}{\omega^2-f^2} = \left[\frac{\pi n}{(C_1+2D)^{1/2} - (C_1)^{1/2}} \right]^2$$

If the constant C_1 is large enough so that $C_1 >> 2z$ for all z, then $(C_1-2z)^{1/2} \doteq (C_1)^{1/2} \doteq (C_1)^{1/2}\left(1 + \frac{z}{C_1}\right)$, and the eigensolution are given by

$$\mathcal{W}_n(z) = a_n \sin\frac{n\pi z}{D}$$

as in the case when N is constant [Eq. (20)].

Other continuous density profiles

The Brunt-Väisälä frequency, N, may be considered as a random function of time, as well as of depth (Barcilon *et al.*, 1972; Bell, 1973c;

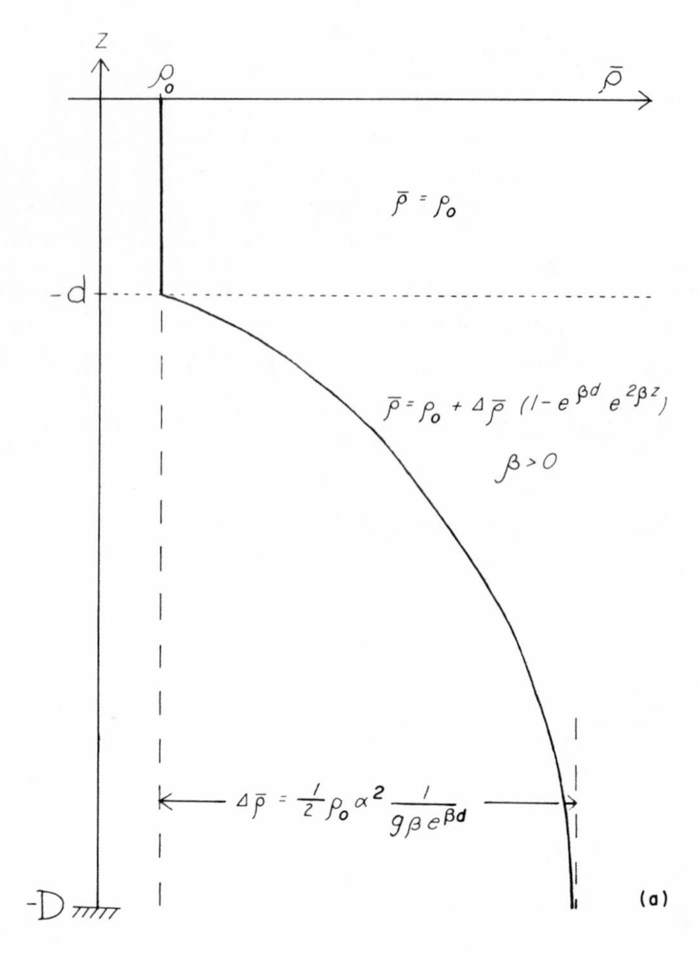

FIG. 3-17. (a) Density and (b) Brunt-Väisälä frequency for the case of a homogeneous layer overlying a layer where N is exponential.

Covez, 1971; Liu, 1970; McGorman, 1972; McGorman and Mysak, 1973; Mysak, 1973). McGorman (1972) has obtained a dispersion relation for internal waves when N is random.

Krauss (1966b, pp. 40-41) has discussed eigensolutions to Eq. (11) in the case of long waves when N(z) may be represented by a series expansion.

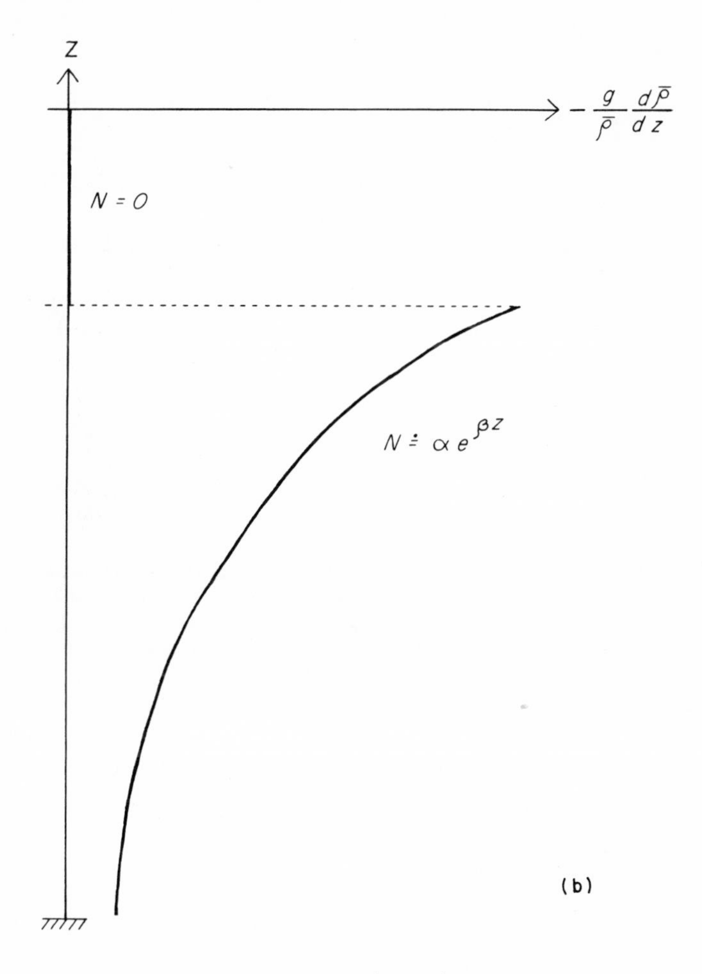

FIG. 3-17 (cont.).

Cox and Sandstrom (1962) have treated the model

$$N^2 = \begin{cases} 0 & 0 \geqslant z \geqslant -d \\ -\dfrac{\beta g}{z} & -d \geqslant z \geqslant -D \end{cases} \tag{48}$$

The eigenfunctions are difficult to calculate for this case, and Cox and Sandstrom have plotted only $\mathcal{W}_1$ and $\mathcal{W}_2$.

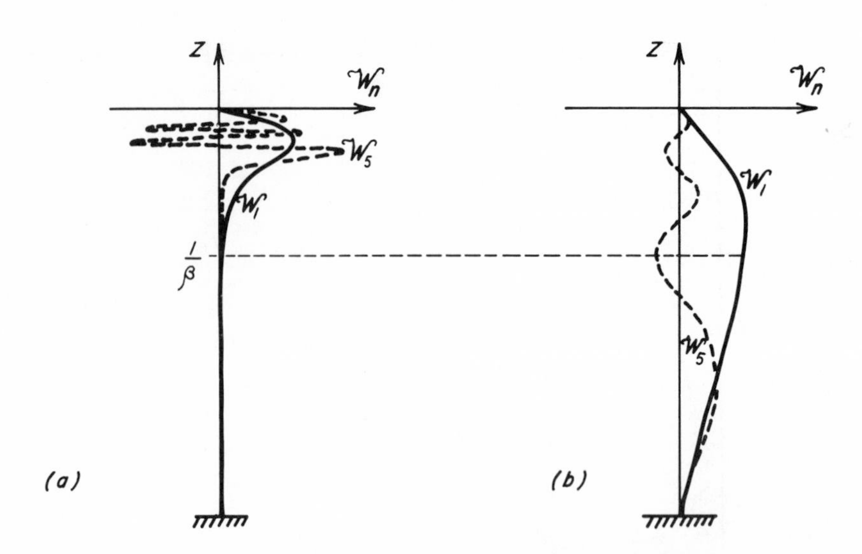

FIG. 3-18. The first and fifth eigensolutions for the case of a homogeneous layer overlying a layer where $N \doteq \alpha \exp(\beta z)$; (a) $\omega/N = 0.75$; (b) $\omega/N = 0.03$. (Redrawn from Garrett and Munk, 1972b. Garrett and Munk show the dimensionless vertical displacement, which is proportional to the vertical velocity.)

A three-fluid system in which the upper (infinite) layer is homogeneous, the middle layer (the thermocline) is of finite thickness and has constant Brunt-Väisälä frequency N_1, and the lower (infinite) layer has constant Brunt-Väisälä frequency N_2, has been used by Sabinin (1966) to explain why internal waves of frequency N_2 often occur in the thermocline region.

The dispersion relation for the case of two thermoclines has been discussed by Eckart (1961).

Two-fluid system: interfacial waves (Fig. 3-22)

Because of its simplicity, this model of the ocean is used quite often, particularly in earlier literature. In this case, we have only one internal wave mode (Yih, 1960, p. 491); its eigenfunction [neglecting $(N^2/g)(d\mathcal{W}/dz)$] is a solution of Eq. (11) and is given by

$$\mathcal{W}(z) = \begin{cases} C_1 \cosh(\omega\sqrt{\nu}z) + C_2 \sinh(\omega\sqrt{\nu}z) & 0 \geqslant z \geqslant -d \\ C_3 \cosh(\omega\sqrt{\nu}z) + C_4 \sinh(\omega\sqrt{\nu}z) & -d \geqslant z \geqslant -D \end{cases} \tag{49}$$

The eigenvalue ν is given by Eq. (7). The constants C_1, C_2, C_3, and C_4 must be chosen so that the eigenfunction agrees at the boundary $z = -d$, and so that $\mathcal{W}(0) = \mathcal{W}(-D) = 0$. It follows immediately that $C_1 \equiv 0$. Krauss (1966b, pp. 44-45) has given the details for determining the remaining constants. When this is done a dispersion relation is obtained.

$$\tanh(\omega\sqrt{\nu}D) = \frac{\Delta\bar{\rho}\left[\sinh(\omega\sqrt{\nu}d)\cosh(\omega\sqrt{\nu}d) + \frac{g\sqrt{\nu}}{\omega}\sinh^2(\omega\sqrt{\nu}d)\right]}{\rho_2\sinh^2(\omega\sqrt{\nu}d) - \rho_1\cosh^2(\omega\sqrt{\nu}d) + \frac{g\sqrt{\nu}}{\omega}\Delta\bar{\rho}\sinh(\omega\sqrt{\nu}d)\cosh(\omega\sqrt{\nu}d)} \tag{50}$$

(Krauss, 1966b, p. 44).

If the earth's rotation is neglected, so that $f = 0$, then the eigenvalue ν becomes

$$\nu = \frac{k_h^2}{\omega^2} \tag{51}$$

and $\omega\sqrt{\nu}$ is simply k_h. If we let $d_1 = d$ and $d_2 = D-d_1$, then $\tanh(\omega\sqrt{\nu}D) = \tanh[(k_h(d_2+d_2)]$, and after some algebraic manipulation, Eq. (50) becomes

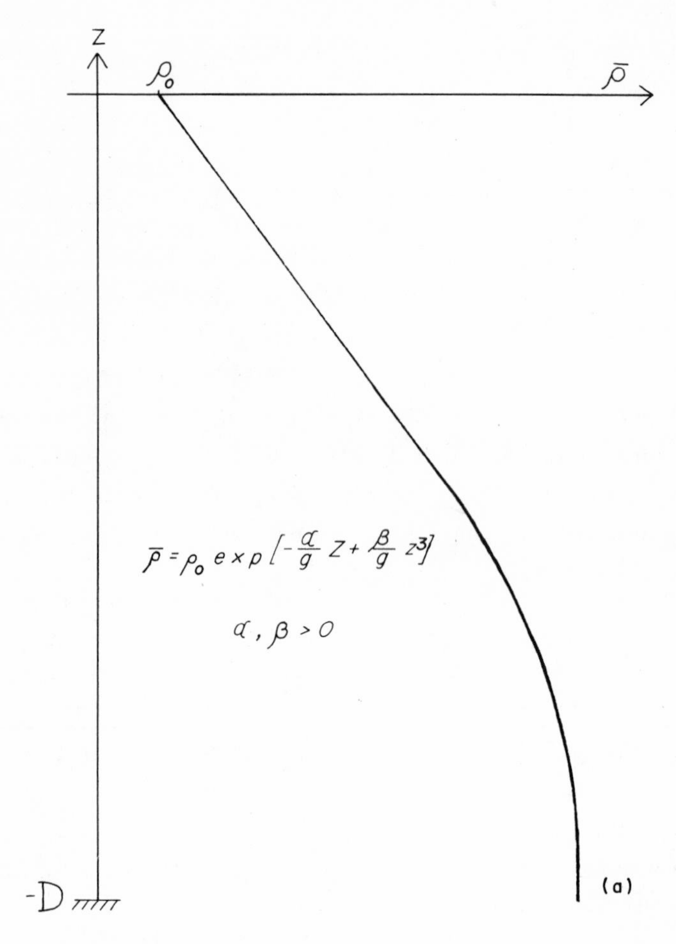

FIG. 3-19. (a) Density and (b) Brunt-Väisälä frequency when the density is given by $\bar{\rho}(z) = \rho_0 \exp[(-\alpha/g)z + (\beta/g)z^3]$; α, $\beta > 0$.

$$\omega^2 - \frac{g\ k_h\ \Delta\bar{\rho}}{\rho_2 \coth(k_h d_2) + \rho_1 \coth(k_h d_1)} = 0 \tag{52}$$

in agreement with Lamb (1945, p. 371).

Lamb (1945, pp. 371-372) and Kinsman (1965, pp. 172-173) have given a discussion of the various approximations which may be made

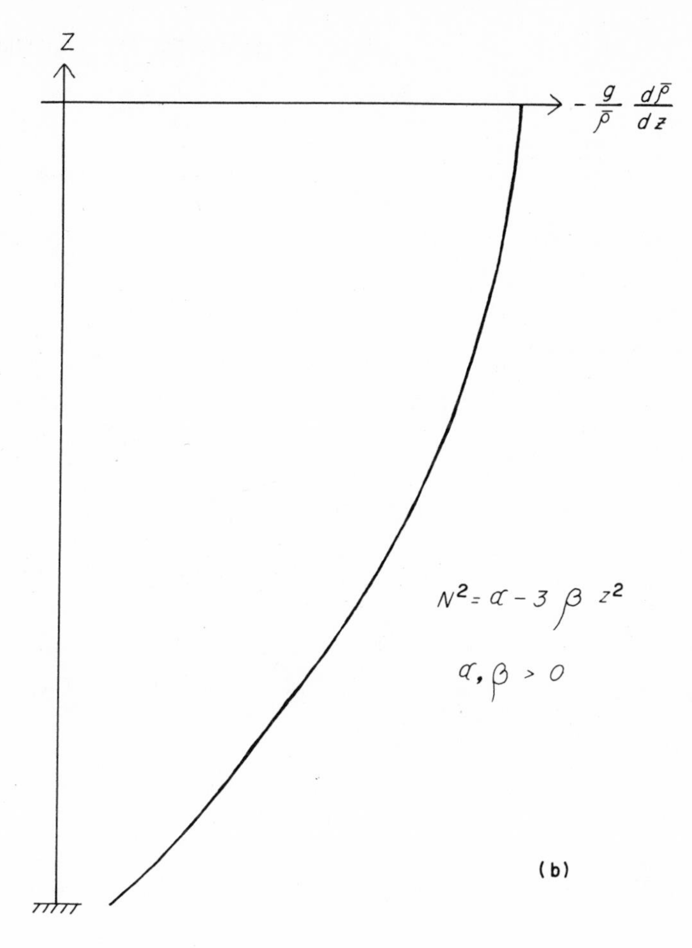

FIG. 3-19 (cont.).

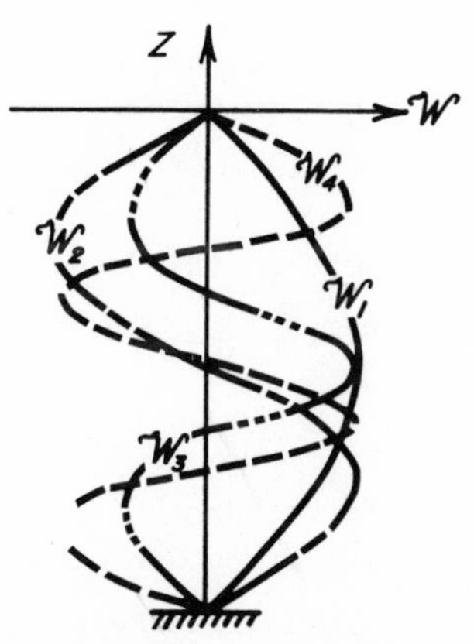

FIG. 3-20. The first four eigensolutions for a system whose density is given by $\bar{\rho}(z) = \rho_0 \exp[(-\alpha/g)z + \beta/g)z^3]$; α, $\beta > 0$.

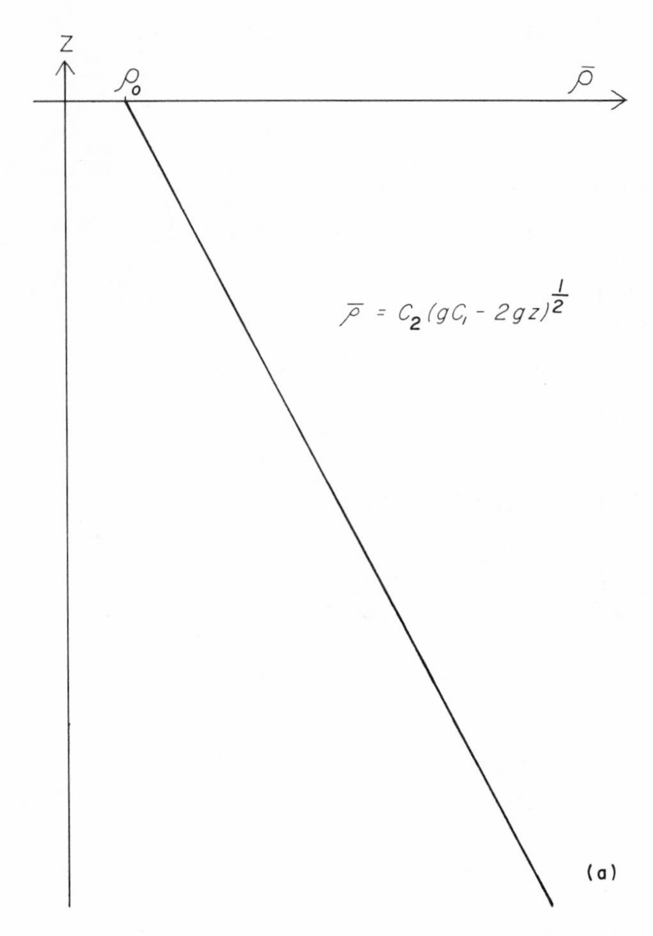

FIG. 3-21. (a) Density and (b) Brunt-Väisälä frequency when the density is given by $\bar{\rho} = C_2(g\ C_1 - 2gz)^{1/2}$. C_1 is assumed to be very large.

to Eq. (52), depending on the depth of the layers d_1 and d_2 relative to the horizontal wavenumber k_h.

Multifluid system: a layered thermocline (Fig. 3-23)

This model assumes that the pycnocline (which in most waters is the thermocline) has a step-like structure made up of a series of *layers* (where the density changes slowly) separated by thin *sheets* (where

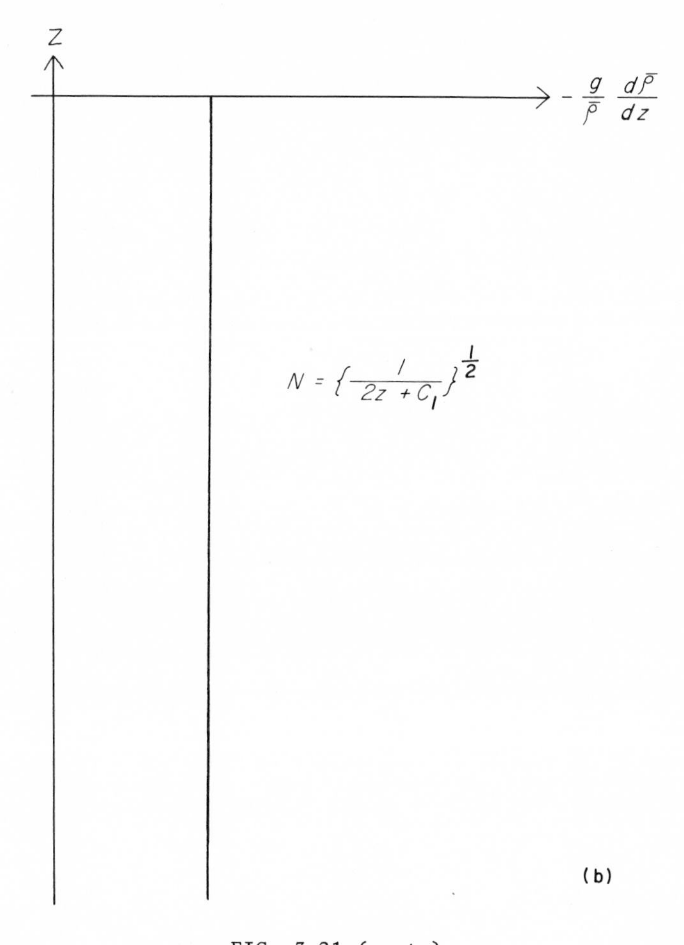

FIG. 3-21 (cont.)

the density changes sharply). This is a good description of the summer thermocline of the Mediterranean (Woods, 1968), and perhaps of most thermoclines. The simplest model assumes homogeneous layers and was adopted by early investigators of internal waves (Greenhill, 1887; Burnside, 1889) to avoid higher transcendental functions. It now appears that this model is at least as realistic as the continuous models we have been examining (Miles, 1972).

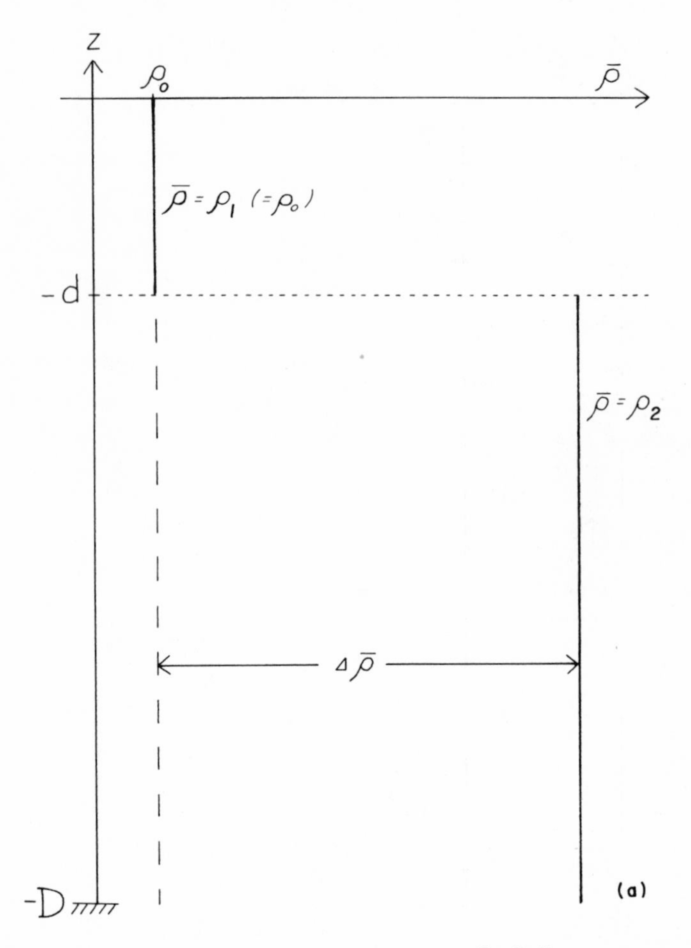

FIG. 3-22. (a) Density and (b) Brunt-Väisälä frequency in a two-fluid system.

In the last section, we saw that a two-fluid system can have internal waves of only first order. It may be noted that in this case both layers contain a zero of the eigenfunction $\mathcal{W}_1$: the top layer contains $\mathcal{W}_1(0) = 0$ and the bottom layer contains $\mathcal{W}_1(-D) = 0$. Yih (1960) has shown that a three-fluid system, in which density changes

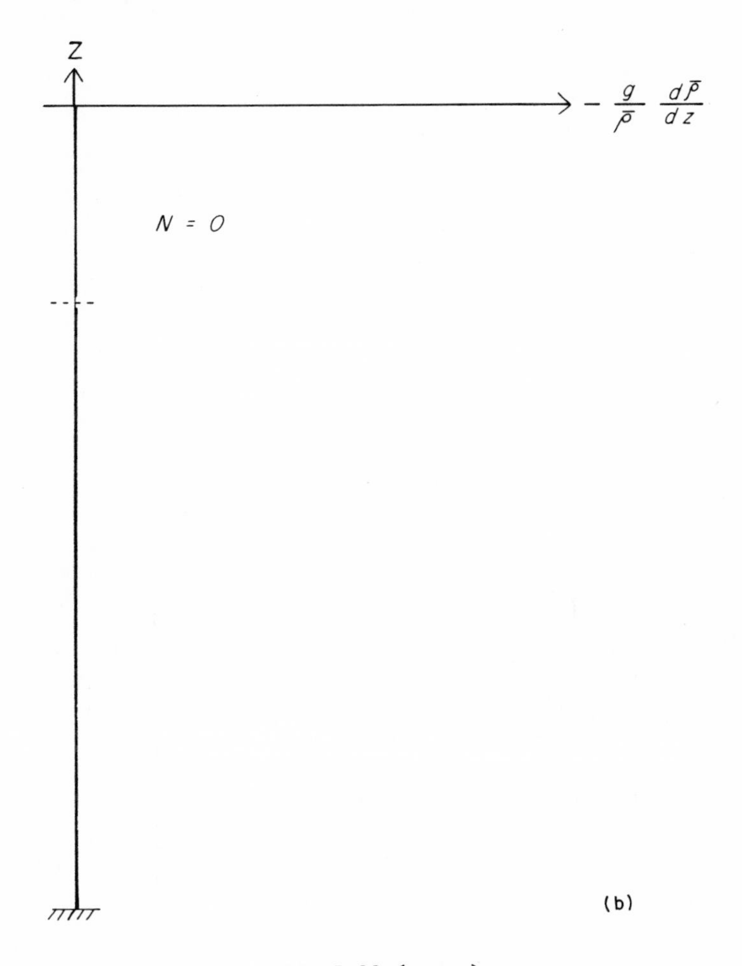

FIG. 3-22 (cont.).

only slightly inside the layers, can have internal waves of only first and second order, and that the middle zero of the eigenfunction $\mathcal{W}_2$ must lie in the middle layer. Thus, each layer has a zero of $\mathcal{W}_2(z)$. The situation is shown schematically in Fig. 3-24. In general, for a system with very small density variations in each layer, zeros in addition to those at the top and bottom will appear one

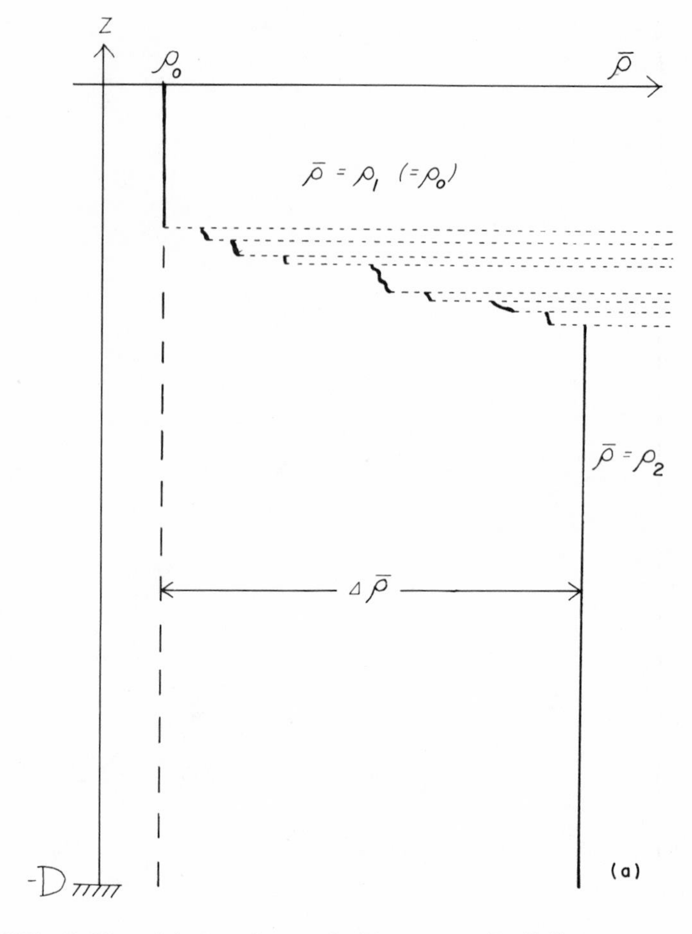

FIG. 3-23. (a) Density and (b) Brunt-Väisälä frequency for a layered thermocline.

after the other as the modes are higher and higher until each layer has one and only one zero. Beyond that, the surfaces of density discontinuity will act like rigid boundaries. Thus, a system with n layers (and hence with n-1 sheets) can support internal waves only of mode up to n-1 (Yih, 1960).

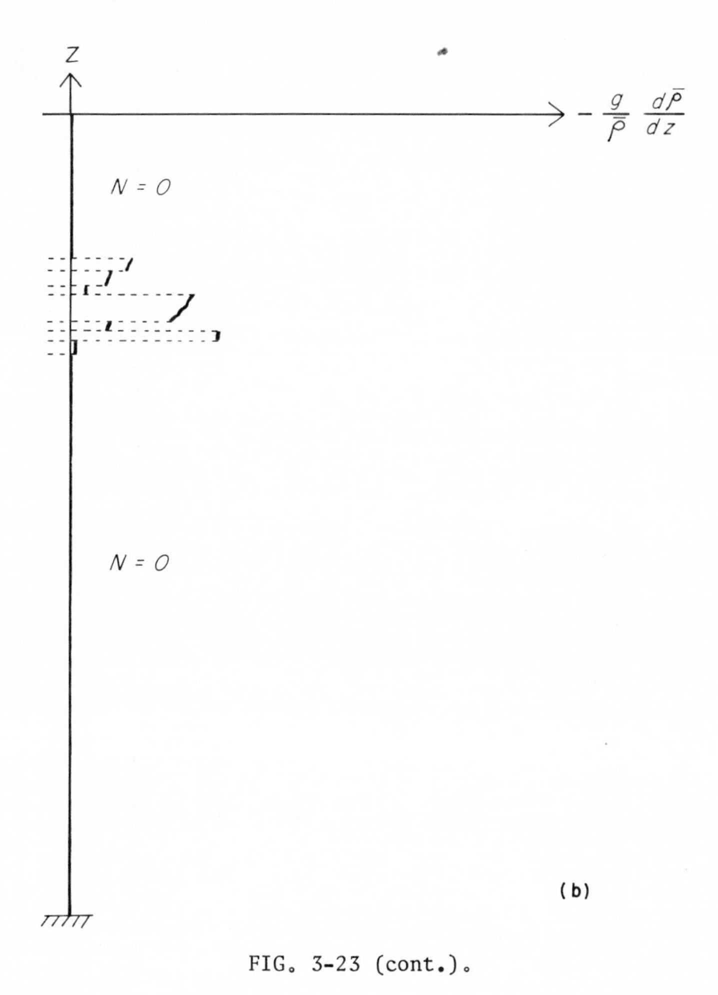

FIG. 3-23 (cont.).

Miles (1972) has posed the eigenvalue problem in the form of a Fredholm integral equation for a multisheeted thermocline. He has found certain bounds for the eigenvalues in the general case of an arbitrary number of sheets separating layers with arbitrary density profiles. The specific cases of three sheets and five sheets, separating layers of constant density, are treated in more detail. (The

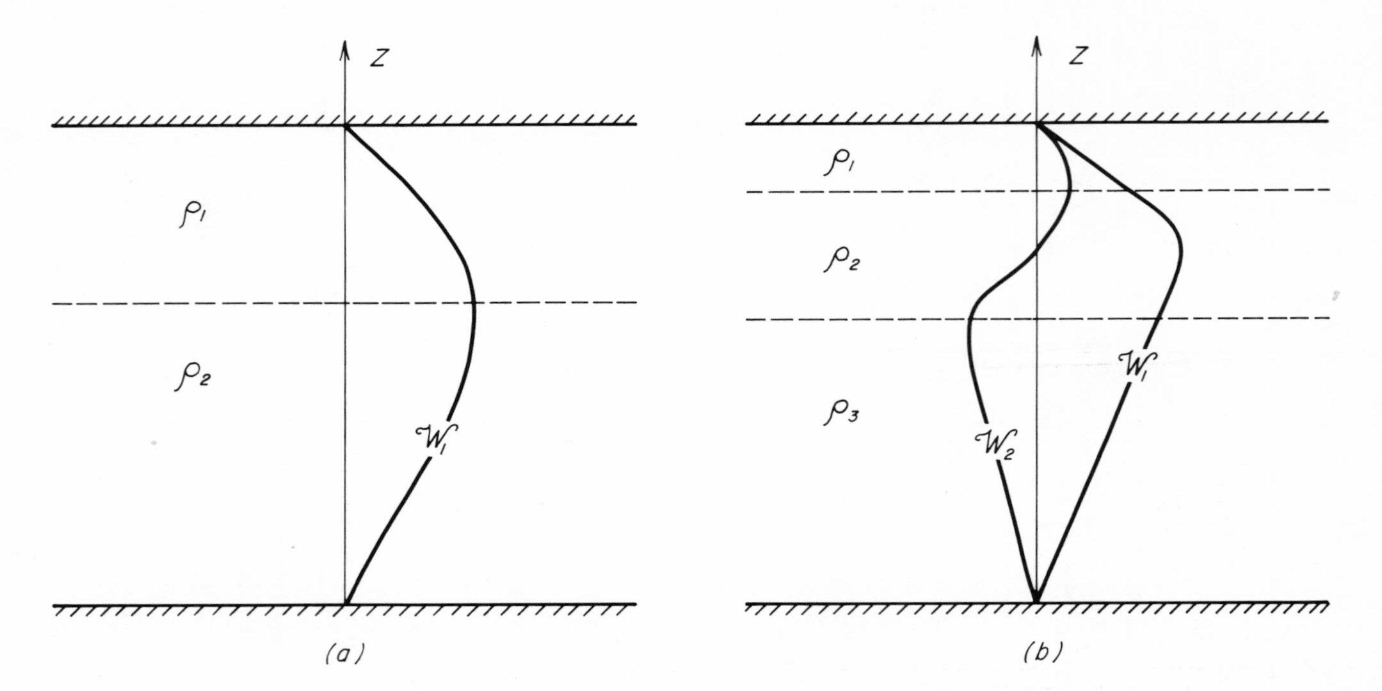

FIG. 3-24. Locations of the zeros (or null points) for the eigenfunction $\mathcal{W}_n$ in (a) a two-fluid system; (b) a three-fluid system. The densities ρ_1, ρ_2, and ρ_3 need not be constant, but should vary little within a given layer.

case of a tanh density is also discussed for comparison.) For a thermocline with three sheets (Fig. 3-25), the eigenvalues are given by

$$\nu_{1,3} = \frac{1}{2}\left\{1 + \nu_2 \pm [(1 - \nu_2)^2 - 4\,\varepsilon\nu_2]^{1/2}\right\} \tag{53a}$$

$$\nu_2 = \frac{1}{2}(1 - \varepsilon)(1 - e^{-2k\ell}) \tag{53b}$$

The layer thickness is given by ℓ. The parameter ε measures the density gradient across a sheet. Since ν is defined by

$$\omega^2 = \nu \frac{gk\,\Delta\bar{\rho}}{\rho_1 + \rho_2} \tag{54}$$

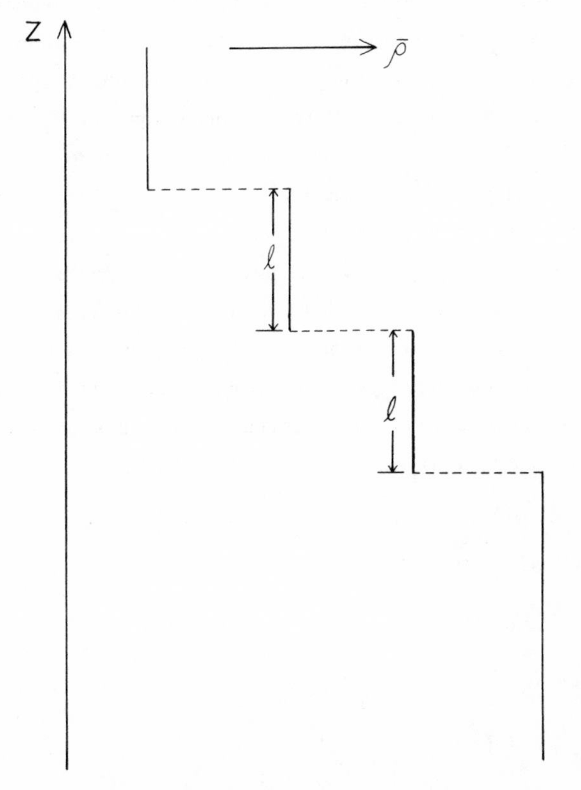

FIG. 3-25. A three-sheeted thermocline with two layers, each of thickness ℓ.

(Fig. 3-23 gives the definition of $\Delta\bar{\rho}$, ρ_1, and ρ_2), Eqs. (53) give a dispersion relation, which Miles has plotted for $\varepsilon = 1/3$, $1/2$, and $2/3$.

The multisheeted model has also been used by Johns and Cross (1969, 1970) to describe the damping of internal waves in the thermocline.

Numerical methods for arbitrary density distributions

The vertical velocity equation, Eq. (8), may be solved numerically using arbitrary values of N(z) determined experimentally. This was first done by Fjeldstad (1933), who used the numerical method of Störmer. His results, however, were valid only for very long waves (Groen, 1948a). Krauss (1966b, pp. 45-46) has discussed the Runge-Kutta numerical integration method applied to the problem. Here, one assumes an initial value $\mathcal{W}_n = 0$ at $z = 0$. Then by trial and error, one obtain eigenvalues ν_n so that the boundary condition $\mathcal{W}_n = 0$ at $z = -D$, holds. Marchuk and Kagan (1970) have used numerical methods to find the eigenvalues and eigenfunctions of the pressure terms, and have given an example for a typical distribution of N. They have found that the high-frequency oscillations are chiefly concentrated in the layer of the principal thermocline, whereas the low frequency oscillations occur in almost the entire mass of the upper ocean where $N > 0$. These results agree with Fig. 3-18.

Documented computer programs for calculating dispersion relations and eigenfunctions for internal wave modes in arbitrary N(z) distributions are available (Bell, 1971; Milder, 1973).

3.5 WAVEFORMS

The waveform for a linear internal wave is described by Eq. (3.4.6a)

$$\nabla_h^2 F + (\omega^2 - f^2)\,\nu F = 0 \tag{1}$$

As before, we set

$$\nu = \frac{k_h^2}{\omega^2 - f^2} \tag{2}$$

where

$$k_h^2 = k_x^2 + k_y^2 \tag{3}$$

and is taken to be positive. k_x and k_y are the wavenumbers in the x- and y-directions, respectively, and are defined by

$$\lambda_x = \frac{2\pi}{k_x} \qquad \lambda_y = \frac{2\pi}{k_y} \tag{4}$$

where λ_x and λ_y are the wavelengths in the x- and y-directions, respectively.

We assume $\omega > f$, so that $\nu > 0$, and apply the method of separation of variables to Eq. (1) as follows:

Suppose that $F(x,y) = aX(x)Y(y)$. Then Eq. (1) becomes

$$\frac{1}{X}\frac{d^2X}{dx^2} + \frac{1}{Y}\frac{d^2Y}{dy^2} + k_h^2 = 0$$

so that

$$\frac{1}{X}\frac{d^2X}{dx^2} = -\frac{1}{Y}\frac{d^2Y}{dy^2} - k_h^2 = -k_x^2$$

where $k_x^2 > 0$. That is,

$$\frac{d^2X}{dx^2} + k_x^2 X = 0 \qquad \frac{d^2Y}{dy^2} - (k_x^2 - k_h^2)Y = 0$$

or letting $k_h^2 - k_x^2 = k_y^2$

$$\frac{d^2X}{dx^2} + k_x^2 X = 0 \tag{5}$$

and

$$\frac{d^2Y}{dy^2} + k_y^2 Y = 0 \tag{6}$$

Since $k_x^2 > 0$, Eq. (5) has the solution

$$X = B_1 \sin k_x x + B_2 \cos k_x x \quad (B_1,\ B_2,\ \text{arbitrary constants}) \tag{7}$$

but the solution of Eq. (6) depends upon the sign of k_y^2.

Case I: $k_y^2 < 0$, i.e., $k_x^2 > k_h^2$. In this case,

$$Y = B_3 \exp[(-k_y^2)^{1/2} y] + B_4 \exp[-(-k_y^2)^{1/2} y]$$

$$(B_3,\ B_4,\ \text{arbitrary constants})$$

or, if we require that the wave not grow exponentially in space,

$$Y = B_4 \exp[-(-k_y^2)^{1/2} y] = B_4 \exp[-(k_x^2 - k_h^2)^{1/2} y] \tag{8}$$

so that

$$F(x,y) = a \exp[-(k_x^2 - k_h^2)^{1/2} y] \begin{Bmatrix} \sin(k_x x) \\ \cos(k_x x) \end{Bmatrix} \tag{9}$$

$$(a,\ \text{an arbitrary constant})$$

Equation (9) describes a Kelvin wave (a type of edge wave) whose crest decreases exponentially in the y-direction (Fig. 3-26).

Case II: $k_y^2 = 0$, i.e., $k_x^2 = k_h^2$. In this case, the wave has no y-dependence, and its crest is of infinite length (Fig. 3-27).

Case III: $k_x^2 > 0$, i.e., $k_x^2 < k_h^2$. In this case

$$Y = B_3 \sin(k_y y) + B_4 \cos(k_y y) \quad (B_3, B_4,\ \text{arbitrary constants})$$

so that

$$F(x,y) = a \begin{Bmatrix} \sin(k_y y) \\ \cos(k_y y) \end{Bmatrix} \begin{Bmatrix} \sin\ (k_x x) \\ \cos\ (k_x x) \end{Bmatrix} \tag{10}$$

(Fig. 3-28).

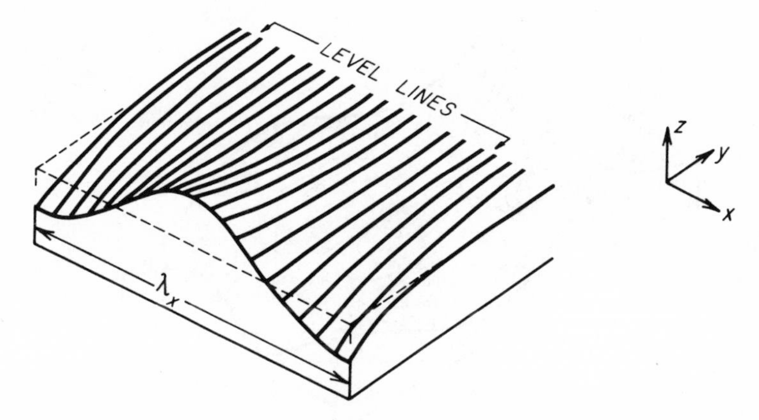

FIG. 3-26. Internal Kelvin wave traveling along one side of a semi-infinite ocean. (Adapted from Mortimer, 1971.)

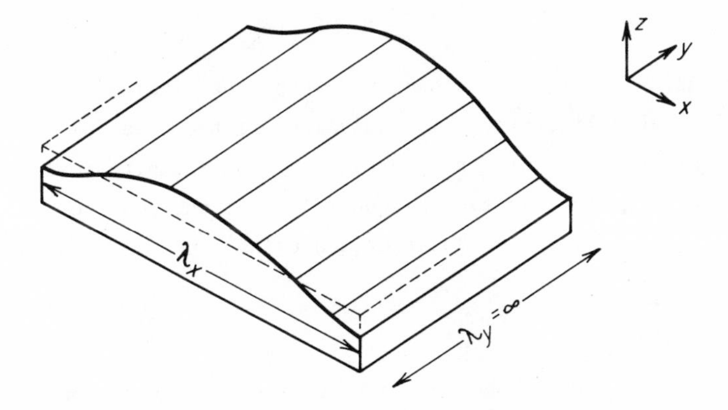

FIG. 3-27. Internal wave with an infinitely long crest.

The above development is due to Krauss (1966b, p. 20 ff.) who also has discussed the cases $\omega = f$ and $\omega < f$.

If the nonlinear equations of motion are developed, the resulting wave will have finite amplitude, and the shape may be other than sinusoidal. Thorpe (1968c) has given an excellent summary of work done up to 1968 on the shape of progressive nonlinear internal waves. A description of his tank experiments and many photographs are included. Magaard (1965) has developed the nonlinear equations of

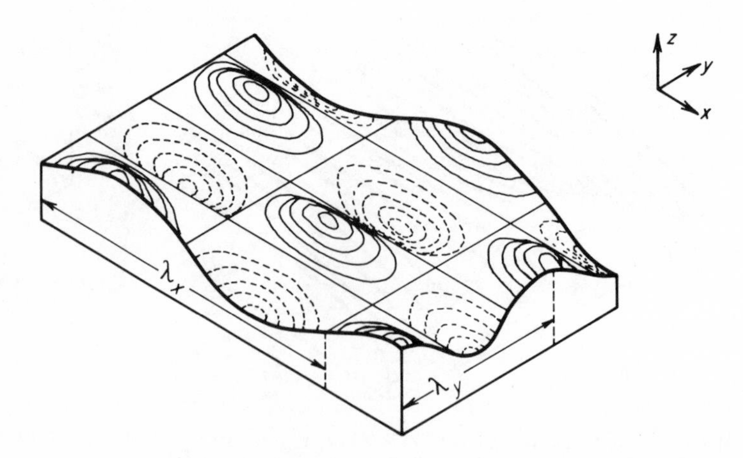

FIG. 3-28. Cellular internal waves. (Adapted from Mortimer, 1971.)

motion using a streamfunction ψ. He has shown that his resulting equation, when linearized, is equivalent to the equation (formulated with a streamfunction) for internal waves of infinitesimal amplitude, but he was unable to solve the general equation. However, by taking the phase speed c_p to be constant and using a modified streamfunction ϕ' defined by:

$$\phi'(s,z) = \psi(s,z) - c_p z$$

where $s = x - c_p t$, he obtained a simplified wave equation

$$\left\{ g \frac{\partial \phi'}{\partial s} + \frac{1}{2} J\left[\left(\frac{\partial \phi'}{\partial s}\right)^2 + \left(\frac{\partial \phi'}{\partial z}\right)^2, \phi'\right]\right\} \frac{1}{\rho} \frac{d\rho}{d\phi'} - J(\Delta\phi', \phi') = 0 \tag{11}$$

where $\rho = \rho(\phi')$ and J denotes the Jacobian of the transformation. Working with this equation, he found an expression for ρ_n, where the n comes from considering eigensolutions ϕ_n, of Eq. (11), and finally obtained an expression for the vertical displacement ζ:

$$\zeta(s,z_o) = \left(\frac{\rho}{d\bar{\rho}/dz}\right) \sin(ks) \sin\left\{\frac{\pi}{D}[z_o - \zeta(s,z_o)]\right\} \tag{12}$$

The extrema of $\zeta(s,z)$ have the same abscissa as the extrema of sin(ks), and therefore Magaard has found that wavelength is independent of the amplitude for the case considered. By taking the wavenumber k to be constant and $\rho = 0.3\Delta\bar{\rho}$, Eq. (12) has been graphed in Fig. 3-29.

The waveform may also be described by writing the amplitude as a perturbation series and considering higher order terms. If we compare surface and internal waves whose first-order amplitudes and wavelengths are equal, we find that for long waves, second-order terms are much greater for internal waves, while for short waves, second-order terms are much greater for surface waves. The conclusion is that short internal waves will retain their linear, sinusoidal shape more easily than short surface waves, while the reverse is true for long internal waves compared to long surface waves (Takano, 1969).

In the last few years, most of the interest in the shape of an internal wave has been concerned with how the wave looks just before it breaks. Some of this work is discussed in Secs. 4.1, 1.4, and 1.9.

3.6 ENERGY AND MOMENTUM CONSIDERATIONS

When we speak of wave energy, we are talking about the energy of a hypothetical particle moved by a wave. The total energy associated with internal waves is usually greater than that associated with surface waves. The following development is due to Krauss (1966b, pp. 46-49).

Consider an nth mode plane internal wave whose vertical velocity is of the form

$$w_n(x,y,z,t) = \mathcal{W}_n(z) \exp[i(k_{x_n} x - \omega t)] \tag{1}$$

where $\mathcal{W}_n$ satisfies Eq. (3.4.8a)

$$\frac{d}{dz}\left[\bar{\rho}(z)\,\frac{d\mathcal{W}_n}{dz}\right] + \nu_n q(z)\,\mathcal{W}_n = 0 \tag{2}$$

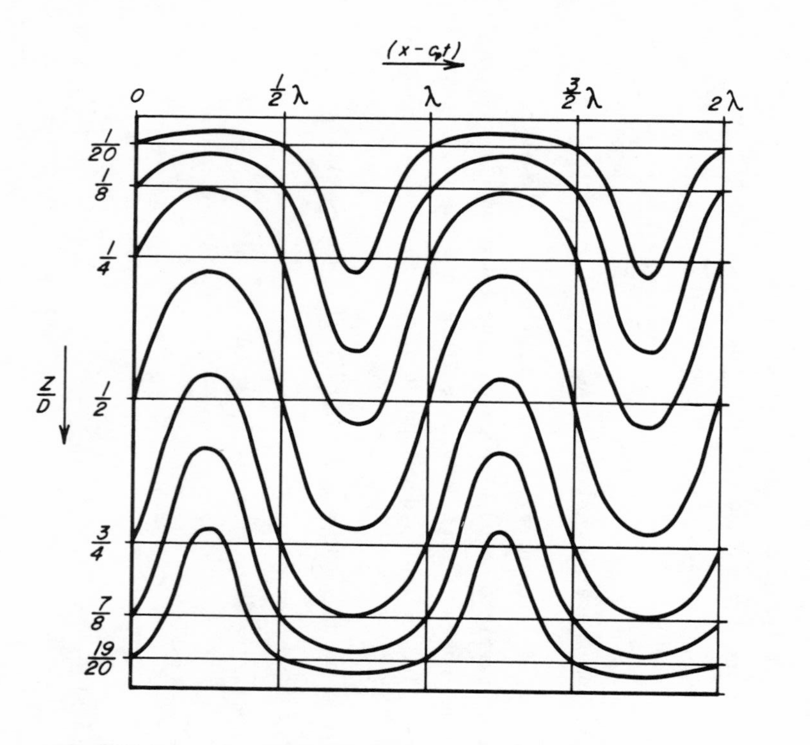

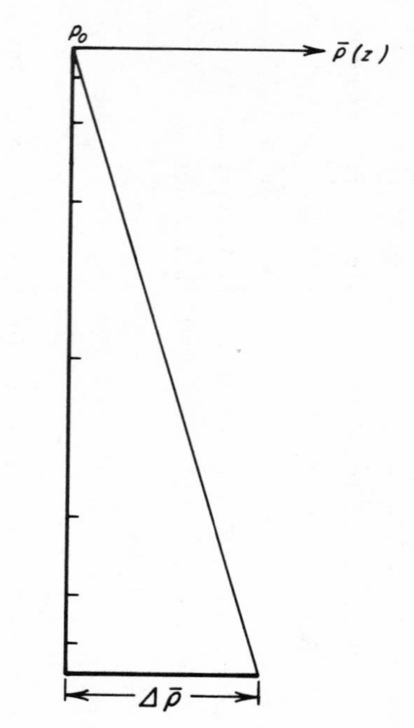

FIG. 3-29. Waveform and density distribution for a finite-amplitude internal wave described by Eq. (1), showing its variation with depth (redrawn from Magaard, 1965). Compare this theoretical picture with the actual profile of isotherm displacement in shallow water, Fig. 1-3.

If we multiply Eq. (2) by the complex conjugate of the vertical velocity amplitude $\mathcal{W}_n^*$, and integrate from 0 to -D, we obtain

$$\int_0^{-D} \mathcal{W}_n^* \frac{d\mathcal{W}_n}{dz}\left(\bar{\rho}\frac{d\mathcal{W}_n}{dz}\right) dz + \nu_n \int_0^{-D} q\left(\mathcal{W}_n^* \mathcal{W}_n\right) dz$$

$$= \mathcal{W}_n^*\left(\bar{\rho}\frac{d\mathcal{W}_n}{dz}\right)\Bigg|_0^{-D} - \int_0^{-D} \bar{\rho}\left(\frac{d\mathcal{W}_n}{dz}\right)\left(\frac{d\mathcal{W}_n^*}{dz}\right) dz$$

$$+ \nu_n \int_0^{-D} q\left(\mathcal{W}_n \mathcal{W}_n^*\right) dz$$

$$= -\int_0^{-D} \bar{\rho}\left|\frac{d\mathcal{W}_n}{dz}\right|^2 dz + \nu_n \int_0^{-D} q\left|\mathcal{W}_n\right|^2 dz$$

$$= 0 \tag{3}$$

From the equation of continuity with $v_n \equiv 0$, we have $\partial u_{x_n}/\partial x = -\partial w_n/\partial z$, so that if $u_{x_n} = \mathcal{U}_n(z) \exp[i(k_{x_n} x - \omega t)]$, then $ik_{x_n}\mathcal{U}_n = -d\mathcal{W}_n/dz$, and

$$\left|\frac{d\mathcal{W}_n}{dz}\right| = k_{x_n}^2 \left|\mathcal{U}_n\right|^2 \tag{4}$$

For $k_{y_n} \equiv 0$ and neglecting the earth's rotation, Eq. (3.4.7) becomes

$$\nu_n q(z) = \frac{N^2(z) - \omega^2}{\omega^2}\bar{\rho}(z)k_{x_n}^2 \tag{5}$$

where $q(z) = [N^2(z) - \omega^2]\bar{\rho}(z)$

Thus, Eq. (3) may be written

$$k_{x_n}^2 \int_0^{-D} \bar{\rho}\left|\mathcal{U}_n\right|^2 dz - \frac{k_{x_n}^2}{\omega^2}\int_0^{-D} (N^2-\omega^2)\bar{\rho}\left|\mathcal{W}_n\right|^2 dz = 0$$

or

$$\int_0^{-D} \bar{\rho}\left(\left|\mathcal{U}_n\right|^2 + \left|\mathcal{W}_n\right|^2\right) dz = \int_0^{-D} \frac{N^2}{\omega^2} \bar{\rho} \left|\mathcal{W}_n\right|^2 dz \tag{6}$$

The left-hand side of Eq. (6) represents twice the kinetic energy of an internal wave, and the right-hand side represents twice the potential energy of an internal wave in a stratified fluid. For a two-fluid system, the potential energy for an infinitesimal wave has the particular simple form $E_{pot} = (1/2)\ g\ \Delta\bar{\rho}\ a^2$, where a is the wave amplitude. The right-hand side of Eq. (6) may be rewritten as follows: The time derivative of the vertical displacement ζ_n, which must also be proportional to $\exp[i(k_{x_n} x - \omega t)]$, is equal to the vertical velocity $w_n \doteq \partial\zeta_n/\partial t$ so that up to a constant

$$\zeta_n \doteq \frac{1}{\omega}\left|\mathcal{W}_n\right| \tag{7}$$

Thus,

$$E_{pot} = \frac{1}{2}\int_0^{-D} \frac{N^2}{\omega^2} \bar{\rho} \left|\mathcal{W}_n\right|^2 dz = \frac{1}{2}\int_0^{-D} N^2\ \bar{\rho} \left|\zeta_n\right|^2 dz \tag{8}$$

(Fofonoff, 1969).

Using Eq. (8), Eq. (6) becomes

$$\frac{1}{2}\int_0^{-D} \bar{\rho}(z) \left(\left|\mathcal{U}_n(z)\right|^2 + \left|\mathcal{W}_n(z)\right|^2\right) dz \quad \text{Kinetic energy of an } n^{th}\text{-order internal wave}$$

$$= \frac{1}{2}\int_0^{-D} N^2(z)\ \bar{\rho}(z) \left|\zeta_n(z)\right|^2 dz \quad \text{Potential energy of an } n^{th}\text{-order internal wave} \tag{9}$$

Thus, we have that for the case of a horizontal ocean floor of depth -D, [so that Eq. (3.3.2) is separable and Eq. (2) holds] and neglecting the earth's rotation, the internal wave energy is equally divided between kinetic and potential energy. This is also Cox's conclusion for the semi-diurnal internal tide (Cox, 1966). The energies of internal waves of different orders n are independent of each other, and in this case there is no energy transferral among waves of different modes. Krauss (1966b, pp. 48-49) has detailed an example of the ratio of energies for different modes using an exponential density distribution (N constant). The maximum amplitude of an internal wave in this case decreases with increasing mode number, and the energy decreases quadratically. Krauss has pointed out that for an internal wave of first order, the total energy may exceed that of a surface wave by as much as 5 to 1.

Internal wave energy, like the energy of most waves, propagates with the group velocity (Whitham, 1961).

When the earth's rotation is taken into account, the kinetic energy need not equal the potential energy and, in fact, generally will not. The ratio of kinetic to potential energy in this more general case has been investigated by Fofonoff (1969), who has compared linear internal wave theory with observed oceanic data. He has concluded that fluctuations of temperature and currents that occur at frequencies between f and N are actually caused by internal waves. Fofonoff's paper contains a valuable appendix on spectral analysis of vector series.

In a stably stratified fluid, lee waves generated by an obstacle transport momentum downstream (Eliassen and Palm, 1961); this momentum transport must be compensated for by a downstream drag, called the *wave drag*, on the obstacle. Most studies of momentum transport and wave drag have been applied to the atmosphere. Many of the results, however, should apply equally well to the ocean. Miles (1968) has given an excellent review of wave drag in stratified flows. More recently, work has been done on the problem by Bretherton (1969), Davis (1969), and McIntyre (1973). Jones (1967) and Bretherton (1969) have considered vertical transport of momentum by internal waves in the atmosphere. Internal waves with periods between 10 min and 2 hr can cause an energy flux out of the troposphere (Gossard,

1962). Jones (1969b) has pointed out that some authors have confused *wave energy flux* with what Jones terms *correlation wave energy flux*. Wave energy flux is associated with E, the wave energy density, a quantity defined (as in Eckart, 1960) in terms of first-order wave perturbations. Thus, if u, for instance, is defined by a perturbation series

$$u = u^{(0)} + u^{(1)} + u^{(2)} + \cdots$$

the wave energy density uses the term $u^{(1)}$. For plane sinusoidal internal waves of amplitude a in a medium at rest

$$E = (1/2)(k_x^2 + k_z^2)a^2 \tag{10}$$

(Bretherton, 1970, p. 84). The correlation wave energy flux, on the other hand, depends on the total energy flux, $\vec{F}_{tot}$, which arises in the equation

$$\frac{\partial}{\partial t}(\rho\ E_{tot}) = -\nabla \cdot \vec{F}_{tot}$$

Here E_{tot} is the total energy density per unit mass, and the equation is valid on an instantaneous basis. With this in mind, the correlation wave energy flux is a result of averaging $\vec{F}$ over one hydrodynamic wave in a fluid.

Energy may be exchanged among waves as a result of (a) resonant interaction, (b) breaking, and (c) reflection and diffraction; in addition, (d) internal waves may exchange energy with ocean currents. These topics are considered in Secs. 2.2, 4.3, 4.4, and 4.7, respectively.

3.7 INERTIAL WAVES

In oceanographic literature, *inertial waves* are defined as horizontal currents in which the current vector rotates with a period T_f of about a half-pendulum day, where

$$T_f = \frac{2\pi}{f} = \frac{\pi}{\Omega \sin \phi}$$

Neumann and Pierson (1966, p. 157 ff.) give the basic equations for a homogeneous ocean, and Phillips (1966, p. 192 ff.) gives a more detailed account for the case N = constant. Webster (1968, 1970) has given an excellent summary of observations of inertial waves.

The frequencies of inertial oscillations in the ocean are almost always greater than the theoretical frequency, f. Theoretical explanations of this phenomenon have been given by Munk and Phillips (1968), and by Tomczak (1968). Gonella (1971) has suggested that the phenomenon may be due to viscosity.

Generation of inertial waves is discussed in Sec. 2.3, and their observations are discussed in Sec. 1.5.

3.8 INTERNAL SEICHES

Just as the surface of a fluid contained in a natural basin, such as a lake, can be deformed periodically by *standing waves* (that is, waves which move the water particles up and down only), so lines of constant density within a stratified fluid in an enclosed basin — the thermocline of a lake, for example — may also exhibit standing waves, or internal seiches. Internal seiches, or temperature seiches as they were first called, have been observed in many lakes and oceanic basins (Sec. 1.5). As was recognized quite early, the main generating mechanism for internal seiches in lakes is wind blowing across the surface and piling up water at one end of the lake. This perturbs the thermocline, which then oscillates with the natural, or seiche, frequency of the basin. Section 1.3 cites a detailed case of wind generation of an internal seiche in Loch Ness.

Theoretically, a standing wave may be described as the sum of two progressive waves of equal amplitude, wavenumber, and frequency, which are traveling in opposite directions. This is most easily seen by considering only the real part of the vertical displacement of a one-dimensional wave,

$$\zeta_1 = Re\left\{a \exp[i(k_x x - \omega t)]\right\} = a \cos(k_x x - \omega t)$$

and

$$\zeta_2 = Re\left\{ a \exp[i(k_x x + \omega t)] \right\} = a \cos(k_x x + \omega t)$$

Then

$$\zeta_1 + \zeta_2 = a[\cos(k_x x - \omega t) + \cos(k_x x + \omega t)]$$

$$= 2a \cos(k_x x) \cos(\omega t)$$

which is the equation of a standing wave. All particles oscillate with the same frequency, but the amplitude of their motion depends upon their position, x. The amplitude of the standing wave has a maximum value of 2a when $k_x x = 0, \pi, 2\pi, 3\pi, \ldots$ or $x = 0, \lambda_x/2, \lambda_x, 3\lambda_x/2, \ldots$; these points are called *antinodes*. The amplitude is zero when $x = \lambda_x/4, 3\lambda_x/4, 5\lambda_x/4, \ldots$; these points are called *nodes*. Because of boundary conditions, the wavelength λ_x is necessarily some fraction of the dimensions of the basin. In a very simplified case for standing waves of a thermocline in a symmetrically enclosed, two-fluid system, the particle motion is analogous to that for a vibrating string fixed at both ends.

When the density varies continuously, internal waves of more than one mode can exist, and the vertical wavelength λ_z must also be taken into account. In this case, it is possible to think of the water as being devided into cells, whose dimensions are defined by the horizontal and vertical wavelengths of the standing wave, and within which oscillation occurs as if between fixed walls (Defant, 1961, pp. 528-532). Figure 3-28 schematically shows the shape of an isopycnal surface deformed by such oscillations. Evidence for cellular oscillations has in fact been found in the Baltic (Krauss, 1966b, pp. 77-79).

Theoretical treatment of standing internal waves in a two-fluid system has been given by Lamb (1945, p. 379); Proudman (1953, pp. 338-340); Kanari (1973); Heaps and Ramsbottom (1966), who included windstress and friction terms; and Thorpe (1968b), who treated standing internal waves of finite amplitude. [Thorpe's paper includes an excellent review of work up to 1967, in particular of theoretical work, on standing internal waves. It also includes an account of a wavetank experiment on the breaking of standing internal waves (Sec. 1.8).] Standing internal waves in a continuously stratified fluid have been discussed theoretically by Krauss (1966b, pp. 73-90); Magaard (1962), who looked at the effect of an uneven bottom topography on the waves; Thorpe (1968b); and Shen (1969). Resonant interaction of standing internal waves has been investigated experimentally and theoretically by McEwan (1971), Joyce (1972), and Orlanski (1972); these results are reported in Sec. 1.8.

3.9 STRATIFIED FLOWS AND LEE WAVES

Stratified flows

The theory of stratified flows includes the theory of internal gravity waves as a special case when the streamlines of the flow are periodic. The stability equation of one-dimensional flows with $V = W = 0$ and $U_x = U_x(z)$ [$=U(z)$] is:

$$\frac{d}{dz}\left[\bar{\rho}(z)\,\frac{d\mathcal{W}}{dz}\right] + \left[\frac{-g\,\frac{d\bar{\rho}}{dz}}{(U-c_p)^2} - \frac{\frac{d}{dz}\left(\bar{\rho}\,\frac{dU}{dz}\right)}{(U-c_p)} - k_h^2\,\bar{\rho}\right]\mathcal{W} = 0 \qquad (1)$$

Here, $\mathcal{W}(z)$ is the amplitude of the vertical velocity w:

$$w(x,y,z,t) = \mathcal{W}(z)F(x,y)e^{-i\omega t} \qquad (2)$$

where F satisfies

$$(\nabla_h^2 + k_h^2)\ F(x,y) = 0 \qquad (3)$$

[Compare with Eqs. (3.4.6a) and (3.4.7).]

Equation (1) may be obtained from linearized versions of Eqs. (3.2.11), (3.2.4), and (3.2.5) (see Yih, 1969, p. 89). If the Boussinesq approximation is made, Eq. (1) becomes, after using Eq. (3.2.16)

$$\frac{d^2\mathcal{W}}{dz^2} + \left[\frac{N^2(z)}{(U-c_p)^2} - \frac{d^2U/dz^2}{(U-c_p)} - k_h^2\right]\mathcal{W} = 0 \qquad (4)$$

which is sometimes called the Taylor-Goldstein equation, because Taylor (1931) and Goldstein (1931) first derived it. Miles (1961) also has given a derivation of Eq. (4). The phase speed c_p is a basic parameter here; if $U = c_p$, the equation is singular, and interesting things can happen physically. Some of these are discussed in Sec. 4.7.

We assume c_p to be made up of a real and an imaginary part

$$c_p = c_r + ic_i \tag{5}$$

We take the wavenumber k to be positive. Then, since

$$\begin{aligned} \omega &= c_p k \\ &= c_r k + ic_i k \end{aligned} \tag{6}$$

the wave represented by Eq. (2) is

$$w = \mathcal{W}(z)F(x,y) \exp(kc_i t) \exp(ic_r kt) \tag{7}$$

You can see that if

$$c_i > 0 \tag{8}$$

the vertical velocity will grow in time, and the flow is said to be *unstable*; kc_i is called the *growth rate*. If $c_i < 0$, the wave will decay.

The Richardson number $N^2(z)/(dU/dz)^2$ is a measure of the amount of shear and is denoted by Ri(z) or J(z). Thus,

$$\text{Ri}(z) = \frac{N^2(z)}{(dU/dz)^2} \tag{9}$$

Miles (1961) and Howard (1961) have formulated the following theorem, known as the Miles-Howard Theorem:

If $\text{Ri}(z) > 1/4$ everywhere, then $c_i = 0$ and the flow is stable. (10)

In a very careful experiment, Scotti and Corcos (1972) have investigated small disturbances in a *free shear layer*, i.e., a transition layer at the common boundary of two streams flowing at different speeds. For the Reynolds number of their experiment ($30 < Re < 70$; $Re \equiv \rho LU/\mu$, where L is a length scale and μ is the viscosity), they found that instability occurred when $Ri < 0.22$, in good agreement with the Miles-Howard theorem.

If the flow is unstable, so that $c_i > 0$, Howard (1961) has shown that the phase speed c_p is bounded by the current speed U(z) in the following manner:

> The complex wave speed c_p for any unstable mode must lie inside the semi-circle in the upper half-plane which has the range of U for diameter. (11)

This is known as Howard's semi-circle theorem and is illustrated in Fig. 3-30 for a given wavenumber k and Richardson number Ri(z).

Howard (1961) also has found the following bound for the growth rate, $k_h c_i$, for an unstable wave:

$$\left[\quad k_h^2 \, c_i^2 \leqslant \max \quad \frac{1}{4}\left(\frac{dU}{dz}\right)^2 - N^2 \right] \tag{12}$$

Growth rates and curves of maximum growth have been numerically investigated by Hazel (1972) for different density and velocity profiles (shear layers, channel profiles, and jet profiles). The first sec-

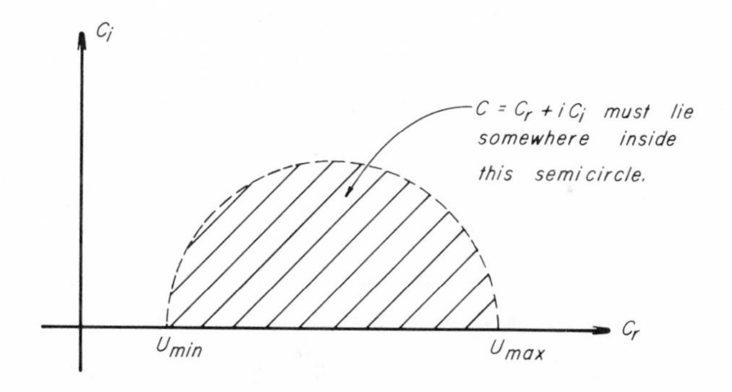

FIG. 3-30. Howard's semi-circle theorem.

tion of Hazel's paper is a very useful introduction to the stability of stratified flows. Scotti and Corcos have experimentally obtained growth rate which are in crude agreement with Hazel's numerical results. The length of the most rapidly growing waves seems to be noticeably affected by relatively minor variations in the mean density and velocity profiles.

Suppose $\mathcal{W}(z) = \mathcal{X}(z) + i\mathcal{Y}(z)$ is a solution to Eq. (4) with eigenvalue $1/c_p = \nu = \nu_r + i\nu_i$ (Eq. 3.4.9); then the complex conjugate, $\mathcal{W}^*(z) = \mathcal{X}(z) - i\mathcal{Y}(z)$ is also a solution with eigenvalue $1/c_p = \nu^* = \nu_r - i\nu_i$. Hence, except when $c_i = 0$, the eigenvalues always appear in pairs, with $c_i > 0$ representing the growing, unstable mode and $c_i <$ representing the decaying mode. The case $c_i = 0$ represents the *neutral mode*. A *singular neutral mode* (SNM) is a neutral mode for which the current speed $U(z_c) = c_r$ at some critical depth z_c, i.e., an SNM occurs when

$$c_i = 0 \qquad \text{and} \qquad U = c_r \tag{12}$$

Since $U(z_o)$ and $\mathcal{W}(z_o)$ are constants at a given depth z_o, for a given c_p, we may write $Ri = Ri(k_h)$ by virtue of Eqs. (4) and (9). Miles (1963) has discussed this functional dependence. In the (k_h, Ri) plane, a *stability boundary* (or *neutral curve*) is a curve made up of SNM's. One of the most important physical aspects of the problem of stability of stratified shear flows is determining the stability boundary of any given configuration in the (k_h, Ri) plane. Hazel (1972) has numerically investigated the stability boundary in the $(k_h Ri)$ plane for various velocity and density profiles, and has given pictures of the constant density surfaces for various flows.

Miles (1963) has mathematically investigated SNM's and shown, among other things, that the existence of an SNM implies the existence of contiguous unstable modes; and that the stability boundary is not, in general, single-valued. He has applied some of these results to the ocean and the atmosphere (Miles, 1967) for the case of the velocity profile

$$U(z) = U'e^{\beta z} \tag{14}$$

and density profile

$$\ln(\rho_0/\bar{\rho}) = \sigma(e^{\beta z}-1) \tag{15}$$

The heterogeneous shear flow thus described is the simplest possible under the restrictions that $U(z)$ and $\ln(\rho_0/\bar{\rho})$ be both bounded and continuous. He has concluded that this flow is stable with respect to small disturbances of all wavelengths and will support an infinite number of internal gravity waves with wave speeds that lie outside the range of the current speeds $(0,U')$, but that the flow will have two limit points: one when c_p is slightly negative and $Ri > 1/4$, and one when c_p is slightly greater than U' and Ri is slightly positive.

Unstable stratified flows are discussed in Chap. 4.

Lee waves

In the atmosphere, waves which occur in the lee of a mountain range standing athwart the prevailing winds often produce striking cloud patterns. Lee waves of geophysical interest, such as those produced by wind blowing over a mountain or tidal currents flowing over a submarine ridge, differ from lee waves on a running stream primarily in that the vertical stratification of the basic flow is gradual rather than abrupt, and secondarily in consequence of *shear* (variation of flow speed with elevation) (Miles, 1968). Most of the work concerning lee waves has been done for atmospheric applications. Miles (1968) has given an excellent review of this work; he includes a development of the equations for two- and three-dimensional flows, a summary of Long's very fine experiments (Long, 1953, 1954a,b), and a discussion of the streamline patterns and wave drag for obstacles of various shapes and sizes. Davis (1969) has pointed out that while there are certain conditions under which wave drag increases with decreasing speed, these conditions also cause the wake to be dominated by turbulence and no lee waves can be detected. Vergeiner (1971) has given a lee wave model for arbitrary basic flow and two-dimensional topography which gives very close agreement with observed stable lee waves. Generally speaking, the basic requirement for stable atmospheric lee waves seems to be moderate winds across the ridge with a wind maximum somewhere at upper levels, but with no extreme wind shears or odd profiles of any kind. Other recent work on lee waves is listed in Sec. 2.4.

In recent years the problem of clear-air turbulence (CAT) which is associated with lee waves, has lent new impetus to their study. It now appears that CAT is caused by breaking internal lee waves (Atlas *et al.*, 1970).

Atmospheric lee waves may be responsible for a significant fraction of the momentum exchange between the earth and the atmosphere; Bell (1973a) has suggested that the same mechanism may transfer energy from currents in the open ocean to internal lee waves. Lee waves in the ocean have been observed near the Straits of Gibraltar (Gade and Ericksen, 1969) and in Massachusetts Bay (Halpern, 1971a), although they are undoubtedly more widespread. Theoretical treatment of the generation and propagation of lee waves in the ocean has been given by Gade and Ericksen (1969), Cavanie (1969, 1971), and Lee (1972). Lee waves behind a moving body are discussed in Sec. 2.5.

3.10 NONLINEAR THEORY

In dealing with internal wave equations, terms such as $u_x(\partial\rho/\partial x)$ or higher-order terms in Taylor's expansion are often ignored. If these quantities are kept in the equations, the development is said to be *nonlinear*. It seems that the linearized equations of motion do an adequate job of explaining internal-wave propagation in the ocean, and that many effects of nonlinearization are secondary (although it should not be concluded that nonlinear effects are therefore unimportant; resonant interaction, for instance, may prove to be very important in the ocean). One of the nonlinear effects is a wave-induced mean flow, proportional to a^2, where a is the wave amplitude (Grimshaw, 1971 and 1973). There may also be an extra term proportional to a^2 added to the pressure term (Drazin, 1969). Rarity (1969) has established conditions under which finite-amplitude waves may propagate, and Weissman (1972) has shown that packets of such waves can travel faster than the group velocity. Shen (1969) has given a mathematical discussion of nonlinear waves which are stationary with respect to a rotating, cylindrical system. A fine development of the nonlinear equations of motion has been given by Magaard (1965).

Nonlinear theory has been used for looking at: (a) the shape of internal waves (Magaard, 1965; Thorpe, 1968c; Benjamin, 1966; Takano, 1969); (b) solitary waves (Benjamin, 1966; Hunkins and Fliegel, 1973); (c) generation of internal waves by long waves over sills (Cavanie, 1969, and 1970; Lee, 1972) and by wave interactions (Joyce, 1972; Hasselmann, 1970); (d) wave instabilities (Drazin, 1970; Maslowe and Kelly, 1970; Smith, 1972; McEwan, 1973); and (e) possible topographic effects (Dore, 1970). Further reference to these works may be found in Secs. 1.10, 1.11, 2.5, 4.3, and 4.6, respectively.

3.11 SOLITARY INTERNAL WAVES

A wave consisting of a single elevation or depression is called a *solitary wave*. Such waves have been observed occasionally in the ocean (Gaul, 1961a; Ziegenbein, 1970; Lee, 1961; Byshev *et al.*, 1971) and in lakes (Hunkins and Fliegel, 1973). For the following discussion, the author is greatly indebted to Thorpe (1968c).

Solitary interfacial waves have been examined by Keulegan (1953, both experimentally and theoretically) and by Long (1956b), both of whom regard the upper surface as fixed. Peters and Stoker (1960) and Walker (1972), have allowed for motion at the upper surface. Whether a solitary interfacial wave is one of elevation or depression depends on the ratios h_1/h_2 and ρ_1/ρ_2 (Fig. 3-31). If $(h_1/h_2)^2 > \rho_1/\rho_2$, the solitary wave will be one of elevation and will resemble a solitary surface wave, while if $(h_1/h_2)^2 < \rho_1/\rho_2$, the solitary wave will be one of depression. If $(h_1/h_2)^2 = \rho_1/\rho_2$, no solitary wave solution is found. The conditions for these results to be valid have been discussed by Thorpe (1968c), Long (1956b), and Hunt (1961). Dyment

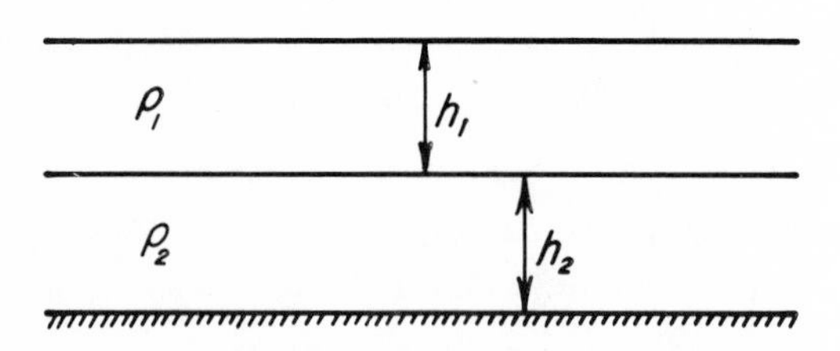

FIG. 3-31. Two-fluid system.

(1968) has also discussed conditions under which a solitary internal wave in a two-fluid system will be one of elevation or depression.

In examining the solitary wave in a fluid with a basic exponential profile, Peters and Stoker (1960) have found solutions for a fluid with a free upper surface, while Long (1965), in a similar fluid but with fixed upper boundary, drew attention to the care which must be taken in making the Boussinesq approximation and the approximation for solitary waves together. The precise nature of these difficulties has been discussed further by Benjamin (1966), who has presented a general treatment for the solution for solitary and cnoidal internal waves in a fluid of great depth. [The waveform of a cnoidal wave can be described as the graph of the square of the Jacobian elliptic function cn x. Kortweg and de Vries, and more recently Benjamin and Lighthill (1954), have discussed cnoidal waves in detail. A solitary wave can be considered as a limiting form of a cnoidal wave (Benjamin and Lighthill, 1954).] Thorpe (1968c) also has considered the consequences of making the Boussinesq approximation for solitary waves. In a later paper, Benjamin (1967) has extended the theory to the case of an infinite fluid in which the density variation is confined to a finite layer, and has examined a new type of solitary wave in this medium in which the vertical scale is established by the layer thickness. The problem has also been studied theoretically and the results well-confirmed experimentally by Davis and Acrivos (1967a). The shape of a solitary internal wave of the first mode in a fluid of exponential density is shown in Fig. 3-32.

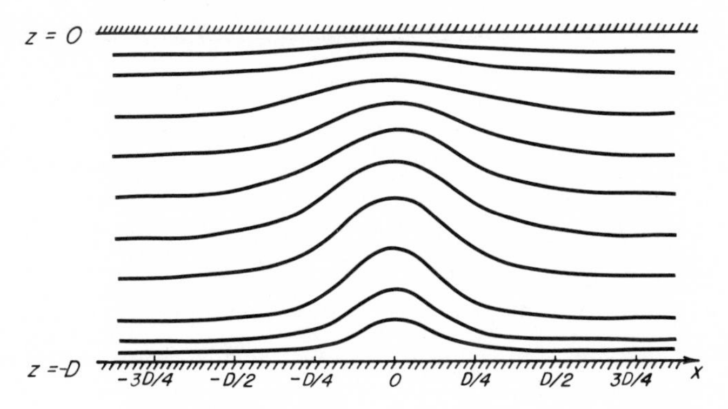

FIG. 3-32. Solitary wave of the first mode in a fluid of density $\rho_0 \exp(-N^2 z)$ between rigid boundaries at $z = 0, -D$. (Redrawn from Thorpe, 1968c.)

Hunkins and Fliegel (1973) have interpreted internal surges in Seneca Lake (which are very similar in appearance to those found by Thorpe *et al.* in Loch Ness, Fig. 1-2) as nonstationary solutions of the Kortweg-de Vries (KDV) equation.

The theoretical existence of internal waves in the atmosphere has been shown by Afashagov (1969) and by Long and Morton (1966). The solitary wave found by Long and Morton depends for its existence on the compressibility of the medium, no matter how small, and its speed of propagation is approximately that of an internal gravity wave.

APPENDIX 3.2.1 The Equivalence of Eq. (3.2.12) and the Equation of Motion Given by Phillips (1966)

If we suppose that the value of $\bar{\rho}(z)$ differs very little from the surface to the bottom of the ocean, that is, if

$$\bar{\rho}(z) \doteq \bar{\rho}(0) \equiv \rho_0, \text{ a constant}$$

and if we suppose that $\vec{F} \equiv 0$ and $\vec{U} \equiv 0$, then Eq. (3.2.12) is equivalent to Phillips' (1966) Eq. (2.4.7). This may be proved as follows:

With the above assumptions, Eq. (3.2.12) becomes

$$\frac{D\vec{u}}{dt} + 2\vec{\Omega} \times \vec{u} + \frac{1}{\rho_0} \nabla p + \hat{k} \frac{g\rho}{\rho_0} = \frac{\mu}{\rho_0} \nabla^2 \vec{u} \tag{1}$$

By the equation of hydrostatics, $\partial\bar{\rho}/\partial z + g\bar{\rho} = 0$, we have

$$\frac{1}{\rho_0} \frac{\partial p}{\partial z} + \frac{g\rho}{\rho_0} = \frac{1}{\rho_0} \left(\frac{\partial \bar{p}}{\partial z} + \frac{\partial p}{\partial z} \right) + \frac{1}{\rho_0} (g\bar{\rho} + g\rho) \tag{2}$$

If we let p_0 represent the hydrostatic pressure in an ocean at rest with constant density ρ_0, then $\partial p_0/\partial z + g\rho_0 = 0$, and Eq. (2) can be written

$$\frac{1}{\rho_0} \frac{\partial p}{\partial z} + \frac{g\rho}{\rho_0} = \frac{1}{\rho_0} \frac{\partial}{\partial z} (\bar{p}+p-p_0) + \frac{1}{\rho_0} (\bar{\rho}+\rho-\rho_0)g$$

$$= \frac{1}{\rho_0} \frac{\partial}{\partial z} (p^*-p_0) + \frac{1}{\rho_0} (\rho^*-\rho_0)g \tag{3}$$

where $p^* = \bar{p}(z) + p(x,y,z,t)$ and $\rho^* = \bar{\rho}(z) + \rho(x,y,x,t)$. Since $p_0 = p_0(z)$, Eq. (1) can be written

$$\frac{D\vec{u}}{Dt} + 2\vec{\Omega} \times \vec{u} + \frac{1}{\rho_0} \nabla(p^*-p_0) + \hat{k} \frac{(\rho^*-\rho_0)}{\rho_0} g = \frac{\mu}{\rho_0} \nabla^2 \vec{u} \tag{4}$$

which is Phillips' Eq. (2.4.7).

APPENDIX 3.2.2. Derivation of Eq. (3.2.14) from Eq. (3.2.12)

The procedure is an elaboration on that given in Krauss (1966b, p. 6 ff.). Equation (3.2.12) may be written

$$\bar{\rho} \frac{\partial \vec{u}^*}{\partial t} + 2\bar{\rho}\vec{\Omega} \times \vec{u}^* + \nabla p + \hat{k}g\rho = \vec{R} \tag{1}$$

where

$$\vec{R} = \mu\nabla^2 u^* - \bar{\rho}(\vec{u}^* \cdot \nabla)\vec{u}^* + \vec{F} = \hat{i}R_x + \hat{j}R_y + \hat{k}R_z$$

Since $\nabla_h \cdot \hat{k}\rho g = 0$, we have

$$\nabla_h \cdot \bar{\rho} \frac{\partial \vec{u}^*}{\partial t} + \nabla_h \cdot (2\bar{\rho}\vec{\Omega} \times \vec{u}^*) + \nabla_h \cdot \nabla p = \nabla_h \cdot \vec{R} \tag{2}$$

Taking the partial derivative of Eq. (2) with respect to time, we have

$$\frac{\partial}{\partial t}\left(\nabla_h \cdot \bar{\rho}\,\frac{\partial \vec{u}^*}{\partial t}\right) + 2\bar{\rho}\nabla_h \cdot \left(\vec{\Omega} \times \frac{\partial \vec{u}^*}{\partial t}\right) + \nabla_h \cdot \nabla\,\frac{\partial p}{\partial t} = \nabla_h \cdot \frac{\partial \vec{R}}{\partial t} \qquad (3)$$

Before proceeding further, we need the relationships given below in Eqs. (4), (5), and (6).

From the equation of continuity, we have

$$\nabla_h \cdot \bar{\rho}\,\frac{\partial \vec{u}^*}{\partial t} = -\,\bar{\rho}\,\frac{\partial^2 w^*}{\partial z \partial t} \qquad (4)$$

Now,

$$\nabla_h \cdot (\vec{\Omega} \times \vec{u}^*) = \frac{\partial}{\partial x}\,(w^*\Omega_y - v^*\Omega_z) + \frac{\partial}{\partial y}\,(u_x^*\Omega_z - w^*\Omega_x)$$

and

$$-\,\vec{\Omega} \cdot (\nabla \times \vec{u}^*) = \nabla \cdot (\vec{\Omega} \times \vec{u}^*) - \vec{u}^* \cdot (\nabla \times \vec{\Omega})$$

$$= \nabla \cdot (\vec{\Omega} \times \vec{u}^*)$$

$$= \frac{\partial}{\partial x}\,(w^*\Omega_y - v^*\Omega_z) + \frac{\partial}{\partial y}\,(u_x^*\Omega_z - w^*\Omega_x) + \frac{\partial}{\partial z}\,(v^*\Omega_x - u_x^*\Omega_y)$$

so that

$$\nabla_h \cdot (\vec{\Omega} \times \vec{u}^*) = -\,\vec{\Omega} \cdot (\nabla \times \vec{u}^*) - \frac{\partial}{\partial z}\,(v^*\Omega_x - u_x^*\Omega_y)$$

and finally,

$$2\bar{\rho}\nabla_h \cdot \left(\vec{\Omega} \times \frac{\partial \vec{u}^*}{\partial t}\right) = -2\bar{\rho}\vec{\Omega} \cdot \left(\nabla \times \frac{\partial \vec{u}^*}{\partial t}\right) - 2\bar{\rho}\left[\frac{\partial^2}{\partial z \partial t}\,(v^*\Omega_x - u_x^*\Omega_y)\right] \qquad (5)$$

Now we take the curl of Eq. (1) and dot the resulting equation with $\vec{\Omega}$, again denoting the right-hand side of Eq. (1) by $\vec{R}$

$$\vec{\Omega} \cdot \left(\nabla \times \bar{\rho}\,\frac{\partial \vec{u}^*}{\partial t}\right) + \vec{\Omega} \cdot [\nabla \times 2\bar{\rho}(\vec{\Omega} \times \vec{u}^*)] + \vec{\Omega} \cdot (\nabla \times \hat{k} g \rho)$$

$$= \bar{\rho}\vec{\Omega} \cdot \left(\nabla \times \frac{\partial \vec{u}^*}{\partial t}\right) - \frac{d\bar{\rho}}{dz}\vec{\Omega} \cdot \left(\hat{i}\,\frac{\partial v^*}{\partial t} - \hat{j}\,\frac{\partial u_x^*}{\partial t}\right) + 2\bar{\rho}\vec{\Omega} \cdot [\nabla\times(\vec{\Omega}\times\vec{u}^*)]$$

$$+ 2\,\frac{d\bar{\rho}}{dz}\vec{\Omega} \cdot [\hat{i}(\Omega_x w^* - \Omega_z u_x^*) + \hat{j}(\Omega_y w^* - \Omega_z v^*)] + \vec{\Omega} \cdot (\nabla\times\hat{k}g\rho)$$

$$= \vec{\Omega} \cdot (\nabla \times \vec{R})$$

Multiplying by -2 and rearranging, we have

$$-2\bar{\rho}\vec{\Omega} \cdot \left(\nabla \times \frac{\partial \vec{u}^*}{\partial t}\right) = 4\bar{\rho}\vec{\Omega} \cdot [\nabla \times (\vec{\Omega}\times\vec{u}^*)] + 2\vec{\Omega} \cdot \Bigg\{\nabla\times\hat{k}g\rho - \nabla\times\vec{R}$$

$$+ \frac{d\bar{\rho}}{dz}\left[\hat{i}\left(\Omega_x w^* - \Omega_z u_x^* - \frac{\partial v}{\partial t}\right) + \hat{j}\left(\Omega_y w^* - \Omega_z v^* + \frac{\partial u_x^*}{\partial t}\right)\right]\Bigg\} \tag{6}$$

Using Eqs. (4), (5), and (6), Eq. (3) becomes

$$-\bar{\rho}\,\frac{\partial^3 w^*}{\partial z \partial t^2} + 4\bar{\rho}\vec{\Omega} \cdot [\nabla \times (\vec{\Omega}\times\vec{u}^*)] + 2\vec{\Omega} \cdot \Bigg\{\nabla\times\hat{k}g\rho - \nabla\times\vec{R}$$

$$+ \frac{d\bar{\rho}}{dz}\left[\hat{i}\left(\Omega_z w^* - \Omega_z u_x^* - \frac{\partial v}{\partial t}\right) + \hat{j}\left(\Omega_y w^* - \Omega_z v^* + \frac{\partial u_x^*}{\partial t}\right)\right]\Bigg\}$$

$$-2\bar{\rho}\,\frac{\partial^2}{\partial z \partial t}\,(v^*\Omega_x - u_x^*\Omega_y) + \nabla_h \cdot \nabla\,\frac{\partial p}{\partial t} = \nabla_h \cdot \frac{\partial \vec{R}}{\partial t}$$

or, upon rearranging and changing signs,

$$\bar{\rho}\,\frac{\partial^3 w^*}{\partial z \partial t^2} - 4\bar{\rho}\vec{\Omega} \cdot [\nabla \times (\vec{\Omega}\times\vec{u}^*)] - \nabla_h^2\,\frac{\partial p}{\partial t} = S \tag{7}$$

where

$$S = 2\vec{\Omega} \cdot \Bigg\{\nabla\times\hat{k}g\rho - \nabla\times\vec{R} + \frac{d\bar{\rho}}{dz}\left[\hat{i}\left(\Omega_x w^* - \Omega_z u_x^* - \frac{\partial v}{\partial t}\right)\right.$$

$$\left.+ \hat{j}\left(\Omega_y w^* - \Omega_z v^* + \frac{\partial u_x^*}{\partial t}\right)\right]\Bigg\}$$

$$- 2\bar{\rho}\,\frac{\partial^2}{\partial z \partial t}\,(v^*\Omega_x - u_x^*\Omega_y) - \nabla_h \cdot \frac{\partial \vec{R}}{\partial t}$$

Now we consider the z-component of Eq. (1)

$$\bar{\rho}\,\frac{\partial w^*}{\partial t} + 2\bar{\rho}\,(v^*\Omega_x - u_x^*\Omega_y) + \frac{\partial p}{\partial z} + g\rho = R_z \tag{8}$$

Differentiating Eq. (8) with respect to time, we get

$$\bar{\rho}\,\frac{\partial^2 w^*}{\partial t^2} + \frac{\partial^2 p}{\partial z\partial t} + g\,\frac{\partial \rho}{\partial t} = \frac{\partial R_z}{\partial t} - 2\bar{\rho}\,\frac{\partial}{\partial t}\,(v^*\Omega_x - u_x^*\Omega_y) \tag{9}$$

From Eqs. (3.2.4) and (3.2.6),

$$\frac{D\rho^*}{Dt} = \frac{\partial \rho}{\partial t} + u_x^*\,\frac{\partial \rho}{\partial x} + v^*\,\frac{\partial \rho}{\partial y} + w^*\,\frac{\partial \rho}{\partial z} + w^*\,\frac{d\bar{\rho}}{dz} = 0$$

so that

$$\frac{\partial \rho}{\partial t} = -\vec{u}^* \cdot \nabla\rho - w^*\,\frac{d\bar{\rho}}{dz}$$

and Eq. (9) becomes

$$\bar{\rho}\,\frac{\partial^2 w^*}{\partial t^2} + \frac{\partial^2 p}{\partial z\partial t} - g(\vec{u}^* \cdot \nabla\rho) - gw^*\,\frac{d\bar{\rho}}{dz}$$

$$= \frac{\partial R_z}{\partial t} - 2\bar{\rho}\,\frac{\partial}{\partial t}\,(v^*\Omega_x - u_x^*\Omega_y) \tag{10}$$

Taking the horizontal Laplacian of Eq. (10) and rearranging, we have

$$-\nabla_h^2\,\frac{\partial^2 p}{\partial z\partial t} = \bar{\rho}\nabla_h^2\,\frac{\partial^2 w^*}{\partial t^2} - g\,\frac{d\bar{\rho}}{dz}\,\nabla_h^2 w^*$$

$$- \nabla_h^2\left[\, g(\vec{u}^*\cdot\nabla\rho) + \frac{\partial R_z}{\partial t} - 2\,\bar{\rho}\,\frac{\partial}{\partial t}\,(v^*\Omega_x - u_x^*\Omega_y)\right] \tag{11}$$

If we differentiate Eq. (7) again with respect to z, we have

$$\bar{\rho}\,\frac{\partial^4 w^*}{\partial z^2\partial t^2} + \frac{d\bar{\rho}}{dz}\,\frac{\partial^3 w^*}{\partial z\partial t^2} - 4\,\frac{d\bar{\rho}}{dz}\,\Omega \cdot [\nabla \times (\vec{\Omega}\times\vec{u}^*)]$$

$$- 4\,\bar{\rho}\,\vec{\Omega}\cdot\left[\nabla \times\left(\vec{\Omega}\times\frac{\partial\vec{u}^*}{\partial z}\right)\right] - \nabla_h^2\,\frac{\partial^2 p}{\partial z\partial t} = \frac{\partial \mathcal{S}}{\partial z} \tag{12}$$

[Observe that if $\bar{\rho}$ is replaced by a constant, ρ_0, in the inertia terms of Eq. (1), then differentiating Eq. (7) with respect to z will make the second and third terms in Eq. (12) disappear.]

By virtue of Eq. (11), dividing through by $\bar{\rho}$, and using $N^2 = -\frac{g}{\bar{\rho}}\frac{d\bar{\rho}}{dz}$, Eq. (12) becomes

$$\nabla^2 \frac{\partial^2 w^*}{\partial t^2} + N^2\nabla_h^2 w^* + 4\frac{N^2}{g}\vec{\Omega}\cdot[\nabla\times(\vec{\Omega}\times\vec{u}^*)]$$

$$- 4\vec{\Omega}\cdot\left[\nabla\times\left(\vec{\Omega}\times\frac{\partial\vec{u}^*}{\partial z}\right)\right] - \frac{N^2}{g}\frac{\partial^3 w^*}{\partial z\partial t^2} = Q \tag{13}$$

where

$$Q = \frac{1}{\bar{\rho}}\frac{\partial S}{\partial z} + \frac{1}{\bar{\rho}}\nabla_h^2\left[g(\vec{u}^*\cdot\nabla\rho) + \frac{\partial R_z}{\partial t} - 2\bar{\rho}\frac{\partial}{\partial t}(v^*\Omega_x - u_x^*\Omega_y)\right] \quad \text{QED}$$

APPENDIX 3.2.3 Proof of Eq. (3.2.21)

We assume

(a) $\nabla\times\vec{u} = \hat{k}\left(\frac{\partial v}{\partial x} - \frac{\partial u_x}{\partial y}\right)$

(b) $\nabla\cdot\vec{u} = 0$

(c) Ω_x, Ω_y, and Ω_z are constant

(d) the order of differentiation can be exchanged for u, v, and w

Then the equality may be shown to hold by direct computation:

$$\vec{\Omega}\cdot\left[\nabla\times\left(\vec{\Omega}\times\frac{\partial\vec{u}}{\partial z}\right)\right] = \Omega_x\frac{\partial}{\partial z}\left(\Omega_x\frac{\partial v}{\partial y} - \Omega_y\frac{\partial u_x}{\partial y} - \Omega_z\frac{\partial u_x}{\partial z} + \Omega_x\frac{\partial w}{\partial z}\right)$$

$$+ \Omega_y\frac{\partial}{\partial z}\left(\Omega_y\frac{\partial w}{\partial z} - \Omega_z\frac{\partial v}{\partial z} - \Omega_x\frac{\partial v}{\partial x} + \Omega_y\frac{\partial u_x}{\partial x}\right)$$

$$+ \Omega_z\frac{\partial}{\partial z}\left(\Omega_z\frac{\partial u_x}{\partial x} - \Omega_x\frac{\partial w}{\partial x} - \Omega_y\frac{\partial w}{\partial y} + \Omega_z\frac{\partial v}{\partial y}\right)$$

$$= \Omega_x^2\left(\frac{\partial^2 v}{\partial y\partial z} + \frac{\partial^2 w}{\partial z^2}\right) + \Omega_y^2\left(\frac{\partial^2 u_x}{\partial x\partial z} + \frac{\partial^2 w}{\partial z^2}\right)$$

$$+ \Omega_z^2 \left(\frac{\partial^2 u_x}{\partial x \partial z} + \frac{\partial^2 v}{\partial y \partial z} \right) + \Omega_x \Omega_y \left(- \frac{\partial^2 u_x}{\partial y \partial z} - \frac{\partial^2 v}{\partial x \partial z} \right)$$

$$+ \Omega_x \Omega_z \left(- \frac{\partial^2 u_x}{\partial z^2} - \frac{\partial^2 w}{\partial x \partial z} \right) + \Omega_y \Omega_z \left(- \frac{\partial^2 v}{\partial z^2} - \frac{\partial^2 w}{\partial y \partial z} \right)$$

Thus,

$$- \vec{\Omega} \cdot \left[\nabla \times \left(\vec{\Omega} \times \frac{\partial \vec{u}}{\partial z} \right) \right] = \Omega_x^2 \left[- \frac{\partial}{\partial x} \left(- \frac{\partial u_x}{\partial z} \right) \right] + \Omega_y^2 \left[- \frac{\partial}{\partial y} \left(- \frac{\partial v}{\partial z} \right) \right]$$

$$+ \Omega_z^2 \left[- \frac{\partial}{\partial z} \left(- \frac{\partial w}{\partial z} \right) \right]$$

$$+ \Omega_x \Omega_y \left[\frac{\partial}{\partial y} \left(\frac{\partial u_x}{\partial z} \right) + \frac{\partial}{\partial x} \left(\frac{\partial v}{\partial z} \right) \right]$$

$$+ \Omega_x \Omega_z \left[\frac{\partial}{\partial z} \left(\frac{\partial u_x}{\partial z} \right) + \frac{\partial}{\partial x} \left(\frac{\partial w}{\partial z} \right) \right]$$

$$+ \Omega_y \Omega_z \left[\frac{\partial}{\partial z} \left(\frac{\partial v}{\partial z} \right) + \frac{\partial}{\partial z} \left(\frac{\partial w}{\partial y} \right) \right]$$

$$= \Omega_x^2 \frac{\partial^2 w}{\partial x^2} + \Omega_y^2 \frac{\partial^2 w}{\partial y^2} + \Omega_z^2 \frac{\partial^2 w}{\partial z^2}$$

$$+ 2\Omega_x \Omega_y \frac{\partial^2 w}{\partial x \partial y} + 2\Omega_x \Omega_z \frac{\partial^2 w}{\partial x \partial z} + 2\Omega_y \Omega_z \frac{\partial^2 w}{\partial y \partial z}$$

$$= (\vec{\Omega} \cdot \nabla)^2 w \qquad \text{QED}$$

APPENDIX 3.2.4 A Direct Derivation of the Linearized Vertical Velocity Equation from Eq. (3.2.13)

The equation which is derived below is Eq. (23) in Sec. 3.2. We proceed directly from Eq. (3.2.13). The procedure is from Phillips (1966, pp. 161-162 and p. 191).

First, we differentiate Eqs. (3.2.13a) and (3.2.13b) by z and add the resulting equations

$$\frac{\partial^2 u_x}{\partial z \partial t} + f\frac{\partial u_x}{\partial z} + \frac{1}{\bar{\rho}}\frac{\partial^2 p}{\partial x \partial z} + \frac{\partial^2 v}{\partial z \partial t} - f\frac{\partial v}{\partial z} + \frac{1}{\bar{\rho}}\frac{\partial^2 p}{\partial y \partial z} = 0 \tag{1}$$

Then differentiating Eq. (3.2.13c) by x and Eq. (3.2.13c) by y, and subtracting the resulting equations from Eq. (1), we have

$$\frac{\partial^2 u_x}{\partial z \partial t} + \frac{\partial^2 v}{\partial z \partial t} - \frac{\partial^2 w}{\partial x \partial t} - \frac{\partial^2 w}{\partial y \partial t} - \frac{g}{\bar{\rho}}\frac{\partial \rho}{\partial x} - \frac{g}{\bar{\rho}}\frac{\partial \rho}{\partial y} - f\left(\frac{\partial v}{\partial z} - \frac{\partial u_x}{\partial z}\right) = 0 \tag{2}$$

Now differentiate Eq. (2) by t

$$\frac{\partial^2}{\partial t^2}\left(\frac{\partial u_x}{\partial z} + \frac{\partial v}{\partial z}\right) - \frac{\partial^2}{\partial t^2}\left(\frac{\partial w}{\partial x} + \frac{\partial w}{\partial y}\right)$$

$$- \frac{g}{\bar{\rho}}\left[\frac{\partial}{\partial x}\left(\frac{\partial \rho}{\partial t}\right) + \frac{\partial}{\partial y}\left(\frac{\partial \rho}{\partial t}\right)\right] - f\frac{\partial}{\partial t}\left(\frac{\partial v}{\partial z} - \frac{\partial u_x}{\partial z}\right) = 0 \tag{3}$$

From Eq. (3.2.13d), we have

$$\frac{\partial \rho}{\partial t} = - w\frac{\partial \bar{\rho}}{\partial z} = - w\frac{d\bar{\rho}}{dz} \tag{4}$$

Substituting Eq. (4) into Eq. (3) and taking the horizontal divergence of the resulting equation, we have

$$\frac{\partial^2}{\partial t^2}\left(\frac{\partial^2 u_x}{\partial x \partial z} + \frac{\partial^2 u_x}{\partial y \partial z} + \frac{\partial^2 v}{\partial x \partial z} + \frac{\partial^2 v}{\partial y \partial z}\right) - \frac{\partial^2}{\partial t^2}\left(\frac{\partial^2 w}{\partial x^2} + 2\frac{\partial^2 w}{\partial x \partial y} + \frac{\partial^2 w}{\partial y^2}\right)$$

$$+ \frac{g}{\bar{\rho}}\left(\frac{\partial^2 w}{\partial x^2} + 2\frac{\partial^2 w}{\partial x \partial y} + \frac{\partial^2 w}{\partial y^2}\right)\frac{d\bar{\rho}}{dz} - f\frac{\partial}{\partial t}\left(\frac{\partial^2 v}{\partial x \partial z} + \frac{\partial^2 v}{\partial y \partial z} - \frac{\partial^2 u_x}{\partial x \partial z} - \frac{\partial^2 u_x}{\partial y \partial z}\right) \tag{5}$$

Ignoring terms which will make Eq. (5) nonlinear, we have

$$\frac{\partial^2}{\partial t^2}\left[\frac{\partial}{\partial z}\left(\frac{\partial u_x}{\partial x} + \frac{\partial v}{\partial y}\right)\right] - \frac{\partial^2}{\partial t^2}\left(\frac{\partial^2 w}{\partial x^2} + \frac{\partial^2 w}{\partial y^2}\right)$$

$$+\frac{g}{\bar{\rho}}\frac{d\bar{\rho}}{dz}\left(\frac{\partial^2 w}{\partial x^2}+\frac{\partial^2 w}{\partial y^2}\right)-f\frac{\partial}{\partial z}\frac{\partial}{\partial t}\left(\frac{\partial v}{\partial x}-\frac{\partial u_x}{\partial y}\right)=0 \qquad (6)$$

Now $\partial v/\partial x - \partial u_x/\partial y$ is the vertical component of vorticity, and it may be shown (Phillips, 1966, p. 191) that

$$\frac{\partial}{\partial t}\left(\frac{\partial v}{\partial x}-\frac{\partial u_x}{\partial y}\right)=f\frac{\partial w}{\partial z}$$

From the equation of continuity, we have

$$\frac{\partial u_x}{\partial x}+\frac{\partial v}{\partial y}=-\frac{\partial w}{\partial z}$$

By definition,

$$N^2=-\frac{g}{\bar{\rho}}\frac{d\bar{\rho}}{dz}$$

Thus, Eq. (6) finally becomes

$$\frac{\partial^2}{\partial t^2}\nabla^2 w+N^2\nabla_h^2 w+f^2\frac{\partial^2 w}{\partial z^2}=0 \qquad (7)$$

which is Eq. (3.2.23). QED

CHAPTER 4

SECONDARY EFFECTS

4.1 INTRODUCTION

As the propagation problem becomes better understood, investigators are turning their attention to the more subtle problems of wave instability, damping, the effects of ocean currents, and microstructure.

It seems reasonable to suppose that after the amplitude of an internal wave reaches some critical amplitude it will break, causing turbulence and mixing, in analogy with surface waves. This appears to be roughly true, although the analogy must be used with caution. For instance, it appears that an internal wave may break backwards as well as forwards, and it is not necessary that an internal wave reach a critical amplitude before it breaks. In fact, a noninfinitesimal amplitude, that is, an amplitude such that the ratio is $a:\lambda$ is not small, is *per se* neither necessary nor sufficient to cause an internal wave to break down (Rarity, 1969).

Many of the theories presented in this chapter have been experimentally verified in wavetanks. Figures for some of the results are shown in Sec. 1.8.

4.2 EFFECTS OF RESONANCE

While it is possible that internal waves in the ocean may become resonantly unstable, experimental evidence thus far has been obtained only in wavetanks. The notation in this section follows that of Sec. 2.2.

Hasselmann (1967) has given a stability criterion which applies to all conservative interactive systems independent of the details of the interaction. The nonlinear interaction between two infinitesimal waves, $(\omega_1, \vec{k}_1)$ and $(\omega_2, \vec{k}_2)$, and a finite wave $(\omega_o, \vec{k}_o)$, where the resonance conditions

$$\omega_o = \omega_1 \pm \omega_2 \qquad\qquad \vec{k}_o = \vec{k}_1 \pm \vec{k}_2 \tag{1}$$

hold, is unstable for the sum interaction and neutrally stable for the difference interaction. That is, if $\omega_o = \omega_1 + \omega_2$ and $\vec{k}_o = \vec{k}_1 + \vec{k}_2$, then the finite wave $(\omega_o, \vec{k}_o)$ will be unstable. If we consider all frequencies to be positive, the unstable finite sum wave will thus have the highest frequency.

The stability of progressive internal waves in a fluid with a thermocline density profile, i.e., in a two-fluid system with a diffuse interface, has been investigated theoretically and experimentally by Davis and Acrivos (1967b). They have considered a single primary wave $(\omega_o, \vec{k}_o)$ of finite amplitude and a pair of free infinitesimal waves, $(\omega_1, \vec{k}_1)$ and $(\omega_2, \vec{k}_2)$ forming a resonant triad with the primary wave. Taking $\omega_o = \omega_1 - \omega_2$ and $\vec{k}_o = \vec{k}_1 - \vec{k}_2$, they have found that if the product $\alpha_1\alpha_2$ is positive, where α_1 and α_2 are defined by

$$\alpha_1 \equiv \frac{1}{a_o a_2}\frac{da_1}{dt} \qquad \text{and} \qquad \alpha_2 \equiv \frac{1}{a_o a_1}\frac{da_2}{dt} \tag{2}$$

then the primary wave $(\omega_o, \vec{k}_o)$ is unstable. They found that positive values of $\alpha_1\alpha_2$ are possible only if the waves $(\omega_1, \vec{k}_1)$ and $(\omega_2, \vec{k}_2)$ are going in opposite directions [in agreement with Hasselmann's criterion, since if the waves $(\omega_1, \vec{k}_1)$ and $(\omega_2, \vec{k}_2)$ are going in opposite directions, ω_1 and ω_2 will have opposite signs, and their difference will have larger absolute value than ω_1 or ω_2], each with mode number greater than one and differing by one. Davis and Acrivos give actual algebraic expressions for the α_i (including

$\alpha_o \equiv \frac{1}{a_1 a_2}\frac{da_1}{dt}$), which involve k_i, ω_i, the amplitude of the stream-function Ψ_i, the density, and the lower and upper boundaries of the fluid. In their wavetank experiments, instability occurred only for sufficiently large values of the primary wave amplitude and for sufficiently short periods. This is because the critical amplitude which they have found is proportional to the wavelength, and decreases as the period decreases (since internal waves with shorter periods have shorter wavelengths). They have confirmed the results of Keulegan and Carpenter (1961), namely, that a wave of fixed amplitude and frequency becomes unstable when the interfacial region becomes thick. If the instability were due to shear, increasing the thickness of the interfacial region would <u>increase</u> the stability (Sec. 4.3). Davis and Acrivos have concluded that for every primary wave there are a countably infinite number of possible disturbance pairs which induce instabilities and transfer energy from the primary wave to waves having different modes, different frequencies, and even different propagation directions.

Progressive internal waves in a fluid system where N varies slowly have been discussed by R. W. Smith (1972). He has shown that for a resonant triad consisting of a long surface wave of period T_o, a short internal wave of period near T_o and wavelength λ_1, and another short internal wave of wavelength near λ_1, (a) the internal wave of period near T_o is unstable, and (b) the two short internal waves can generate a long surface wave.

Standing internal waves in a linearly stratified fluid have been investigated theoretically and experimentally by McEwan (1971 and 1973) and Orlanski (1972). McEwan (1971) has found that if the amplitude of a steadily forced standing internal wave grows larger than a certain critical amplitude (the maximum isopycnal slope for his experimental parameters is about 8°), the wave suffers a progressive and destructive distortion of form. Provided that the forcing is not too vigorous, the process primarily responsible for this is a resonant-interactive instability in which pairs of initially free waves with infinitesimal amplitudes are selectively amplified, in agreement with Davis and Acrivos (1967b). This causes localized formation of sharp density discontinuity layers, which McEwan has called *traumata*. Turbulent disorder is always preceded by the sudden and wide-spread occurrence of these density traumata. They persist and spread, and turbulence appears to result from intensified shear in their vicinity. They are spatially regular and quite repeatable. Figures of McEwan's results are shown in Sec. 1.8. More recently, McEwan (1973) has attempted to isolate experimentally the conditions under which traumata, interpreted as a precursor to wave-induced turbulence, will occur. In this he has been only partially successful. The work does, however, demonstrate the strong nonlinearity in isopycnal distortion which can be brought about by the interaction of two internal waves. Resonant interaction among damped internal waves has been discussed by McEwan *et al.* (1972).

Using perturbation expansion solutions and numerical nonlinear solutions, Orlanski (1972) has found that when generation of turbulence is violent, small eddies start to force a secondary flow characterized by layers of strong jets separated by patches of turbulence. Phillips (1968) has shown another effect of resonance. In an unbounded fluid, internal waves can be trapped in a layer at a finite depth by periodic, small variations either in the density gradient or in a weak, horizontal, steady current.

The stability of progressive internal waves in a linearly stratified fluid is discussed by Martin, Simmons, and Wunsch (1972); see Sec. 2.2. It appears that internal waves for which $a \cdot k_h > 10^{-2}$, i.e., internal waves for which the ratio $a:\lambda_h$ exceeds $1/(2\pi \times 10^2)$, are always resonantly unstable (Martin 1972: personal communication).

4.3 SHEAR AND KELVIN-HELMHOLTZ INSTABILITIES

One way that *shear* may occur is to have the magnitude of a fluid velocity in any one direction vary within a fluid. In simple shearing motion, for example, planes of fluid slide over one another.

In the ocean, horizontal shear occurs when the magnitude of a current varies with depth; there are regions of strong, persistent shear even in the open ocean (Pochapsky 1972: personal communication). As outlined below, currents associated with internal wave motion may cause shear quite apart from any shear due to vertical variations in the mean flow. These currents, particularly when combined with the mean flow, may cause shear which is so large that the internal waves break, in turn causing turbulence and mixing. This seems to be one mechanism by which clear air turbulence or *CAT*, and deep oceanic turbulence are formed (Phillips, 1967; Orlanski, 1971). In a recent paper on oceanic mixing by breaking internal waves, Garrett and Munk (1972a) have concluded that mixing is most likely caused by internal waves breaking due to shear instability.

Classical *Kelvin-Helmholtz instability*, first studied by Kelvin in 1871, is the instability of the steady flow of two deep immiscible fluids, the less dense fluid above, each moving in the same direction parallel to the interface and each having constant velocity and density. For sufficiently small differences in the velocities above and below the interface, there are no growing wave-like disturbances of any wavenumber. As the velocity difference, ΔU_x ($=\Delta U$, since $V = W \equiv 0$) is increased, first one wavenumber k_c and then a whole range of wavenumbers including k_c become instable. It was found that the critical wavenumber k_c is given by

$$k_c = [g(\rho_2-\rho_1)/\gamma]^{1/2} \tag{1}$$

The steady velocity difference ΔU_c at which instability first occurs is

$$\Delta U_c = \left\{2\,\frac{\rho_1+\rho_2}{\rho_1\rho_2}\,[g(\rho_2-\rho_1)\gamma]^{1/2}\right\}^{1/2} \tag{2}$$

where ρ_1 is the density of the upper fluid, ρ_2 that of the lower fluid, and γ is the interfacial surface tension (Thorpe, 1969a).

Lately, as Weissman (1972) pointed out in a talk before the American Geophysical Union, the term "Kelvin-Helmholtz instability" has been applied to thin, but continuous, shear layers in stratified fluids. "This is unfortunate," Weissman said, "for two reasons. First, neither Kelvin nor Helmholtz considered continuous profiles; that achievement was Rayleigh's. Second, the continuous case is stabilized differently from the discontinuous case." In Kelvin's model, short waves are stabilized by surface tension or, if surface tension is absent, short waves are never stabilized, as shown by Eq. (2). However, if there is a finite thickness, no matter how small, to the shear layer, short waves are always stabilized when their wavelength becomes less than five or six times the thickness of the layer. Because of this semantic confusion, any instability which results from a velocity gradient within a fluid, discontinuously or continuously stratified, is referred to simply as *shear instability* in the following discussion.

Long (1956a, 1972) has given a theoretical account of the conditions which determine whether a long internal wave will break forward or backward. If the shear across the interface of a two-fluid system is small, the wave breaks forward only if the amplitude is fairly small and the lower layer is thinner than the upper layer. If the upper layer is thinner, the wave breaks backward. If shear exists such that there are higher velocities in the upper layer in the direction of wave propagation, then the forward breaking of a wave is favored. Shear in the opposite direction may cause a wave to break backward. In a fluid with continuous density, the density profile and mode number of the wave influence the direction in which waves break (Long, 1972).

Phillips (1966, pp. 166-168) has considered a system in which a very thin thermocline at depth $z = -d$ lies between two homogeneous fluids. We suppose $\vec{k} = \hat{i}k_x$ and that the three Cartesian velocity components (Sec. 3.2) are

$$u = \mathcal{U}(z)\exp[i(k_x x-\omega t)] \qquad v \equiv 0 \qquad w = \mathcal{W}(z)\exp[i(k_x x-\omega t)] \tag{3}$$

If the vertical displacement of the thermocline is given by

$$\zeta = a\,\exp[i(k_x x-\omega t)] \tag{4}$$

then

$$\left.\frac{d\zeta}{dt}\right|_{z = -d} = w(-d)$$

and

$$\mathcal{W}(-d) = -i\omega a \tag{5}$$

From the equation of continuity (Eq. 3.2.13e)

$$\frac{d\mathcal{W}}{dz} = -ik_x \mathcal{U} \tag{6}$$

Then Eq. (3.4.13) applied at the thermocline gives

$$-ik_x \frac{d\mathcal{U}}{dz} + \left(\frac{N^2(z) - \omega^2}{\omega^2}\right) k_x^2 \mathcal{W}(-d) = 0$$

or

$$\frac{d\mathcal{U}}{dz} = -\left(\frac{N^2(z) - \omega^2}{\omega^2}\right) k_x \omega a \tag{7}$$

Since we are only considering ω such that $\omega^2 < N_m^2$, where N_m^2 is the maximum value of $N^2(z)$, $d\mathcal{U}/dz$ is a maximum at $N_m^2(z)$. At this point, the Richardson number [Eq. (3.9.9)] has the value

$$R_i = \frac{N^2}{(d\mathcal{U}/dz)^2} \tag{8a}$$

or

$$Ri = \frac{N_m^2 \omega^2}{(N_m^2 - \omega^2)^2} \cdot \frac{1}{k_x^2 a^2} \tag{8b}$$

If this local Richardson number falls below some critical value, there will be instability in the wave-induced flow. This criterion for instability has been considered by Hall and Pao (1971) and by Woods (1968). Hall and Pao (1971) have shown that, for internal waves in a two-fluid system incident upon a gently sloping beach, the primary mechanism which causes the waves to break is shear instability. Section 1.8 has a discussion of their experiment, including figures. Davis and Acrivos (1967) remark that the local Richardson number increases with the thickness of the interfacial region, as may be seen from Eq. (8a).

In a series of excellent experiments, Thorpe (1968a, 1969a, 1971b) has been able to reproduce in the laboratory the dye formations observed in the Mediterranean thermocline by Woods (1968) and Woods and Fosberry (1967) (see Fig. 1-8). The process by which the laboratory turbulence is formed has been sketched by Thorpe (1969b) (see Fig. 4-1). In a multilayered system, such as the Mediterranean thermocline, an increase in the number of sheets will cause an increase in the Richardson number, and hence a corresponding increase in wave stability (Johns and Cross, 1970).

In the atmosphere, the energy and momentum removed from the troposphere by the breaking of large amplitude lee waves may be a significant factor in the large-scale circulation. There is some general agreement of the factors that lead to wave breakdown and absorption: large wave amplitude, low Richardson number, decreasing density, and the presence of critical layers. Lilly (1972) has given a summary of the atmospheric problem up to 1971. It appears probable that the mechanisms causing CAT are quantitatively the same as those causing turbulence in the Mediterranean thermocline (Woods, 1969b). Some very interesting radar observations on the development and growth of CAT have been made by Atlas *et al.* (1970), by Atlas and Metcalf (1970), and by Gossard, Jensen, and Richter (1971) (Figs. 1-19 and 1-20). Summaries of work done on CAT have been given by Thorpe (1970) and in a collection of papers edited by Pao and Goldburg (1969).

It appears that nonlinear effects may increase the stability of waves in a shear flow. Drazin (1970) has calculated the nonlinear elevation of the interface between two fluids in shear instability. The unstable mode does not break down into turbulence directly, but becomes a periodic nonlinear wave which Drazin suggests will become steady if the fluid has viscosity, however small. However, the re-

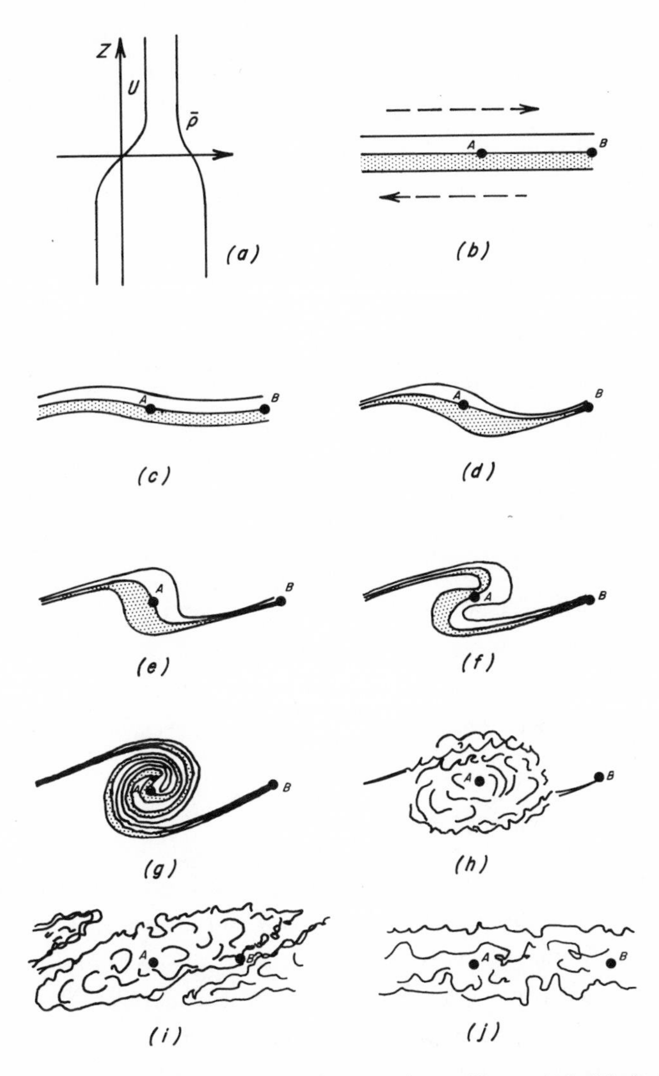

FIG. 4-1. Growth of disturbances in a flow with Ri=0.077±0.01. (a) The density $\bar{\rho}$ and velocity $\vec{U}$ distribution, (b) the lines mark a fluid of constant density, the points A and B are fixed, the arrows indicate the direction of flow. (c) - (j) Development of instability, the lines continue to mark a fluid of constant density. These diagrams are from photographs of laboratory experiments by Thorpe (1968b), and have been redrawn with his permission.

sults are not expected to apply to the development of highly unstable linear waves. In a similar situation, Maslowe and Kelly (1970) have found that periodic finite-amplitude motions are possible for wavenumbers that lie in a region unstable according to linear theory, and have determined a definite lower bound for the equilibrium amplitude of such waves. On the other hand, Weissman (1972) has found that nonlinear instability is possible in a region which is stable in the linear theory. He has also found that packets of large amplitude waves can travel faster than the group velocity.

Hazel (1972) has computed growth rates and curves of maximum growth for various density and velocity profiles. When the scale of the density variation is much less than that of the velocity variation in a shear layer, the flow is unstable for some wavenumber however large the Richardson number may be. Maslowe and Kelly (1971) have also computed stability curves for both spatially and temporally growing disturbances in a shear layer.

A brief discussion of the experiments of Scotti and Corcos (1972) on the stability of free shear layers has already been given in Sec. 3.9. In a recent paper, DeLisi and Corcos (1973) have detailed the density structure of Kelvin-Helmholtz waves as they break.

4.4 TOPOGRAPHIC EFFECTS

From theoretical, numerical, and experimental work, it appears that as internal waves propagate into shallow water, their amplitudes increase while their wavelengths decrease. In addition, there is an increase in particle motion at the bottom boundary on the slope. The methods of geometrical optics are often used to describe various topographic effects on internal waves, so that a very brief review of the idea of rays follows. A more complete discussion with geophysical applications may be found in Eckart (1960, Chap. 11) and Krauss (1966b, Sec. 14). Keller and Mow (1969) have used ray theory to discuss the propagation of internal waves in fluids with variable bottom topography. We have used the W-K-B approximation (Eckart,

1960, pp. 154-156) which, in general, assumes the bottom changes slowly.

Assume an internal wave with vertical velocity of the form

$$w(x,y,z,t) = a \exp[i(k_x x + k_y y + k_z z - \omega t)]$$

where k_z depends on k_x, k_y, ω, and z. For convenience, we use the notation

$$S(k_x, k_y, \omega, z) = k_z z$$

If we superpose two waves of type (1), each of the same amplitude, then (considering a real wave) we get

$$w = a \quad \sin(k'_x x + k'_y y - \omega' t + S') + \sin(k''_x x + k''_y y - \omega'' t + S'') \quad (2)$$

Now let

$$k'_x = k_x + \delta k_x \qquad k'_y = k_y + \delta k_y \qquad \omega' = \omega + \delta\omega$$

$$S' = S(k_x + \delta k_x, k_y + \delta k_y, \omega + \delta\omega, z)$$

$$= S(k_x, k_y, \omega, z) + \frac{\partial S}{\partial k_x}\delta k_x + \frac{\partial S}{\partial k_y}\delta k_y + \frac{\partial S}{\partial \omega}\delta\omega + \cdots$$

and

$$k''_x = k_x - \delta k_x \qquad k''_y = k_y - \delta k_y \qquad \omega'' = \omega - \delta\omega$$

$$S'' = S(k_x - \delta k_x, k_y - \delta k_y, \omega - \delta\omega, z)$$

$$= S(k_x, k_y, \omega, z) - \frac{\partial S}{\partial k_x}\delta k_x - \frac{\partial S}{\partial k_y}\delta k_y - \frac{\partial S}{\partial \omega}\delta\omega - \cdots$$

On neglecting squares and higher powers of the increments, Eq. (2) becomes

$$w = 2a \sin(k_x x + k_y y - \omega t + S) \quad \cos\left[\left(x + \frac{\partial S}{\partial k_x}\right)\delta k_x + \left(y + \frac{\partial S}{\partial k_y}\right)\delta k_y - \left(t - \frac{\partial S}{\partial \omega}\right)\delta\omega\right] \quad (3)$$

This is a modulated sine wave, the modulation being determined by k_x, k_y, ω, δk_x, δk_y, and $\delta\omega$. It will have the constant magnitude

$$\cos\ (x_o \delta k_x + y_o \delta k_y - t_o \delta\omega)$$

for all, x,y,z, and t such that

$$x + \frac{\partial S}{\partial k_x} = x_o = \text{constant} \tag{4a}$$

$$y + \frac{\partial S}{\partial k_y} = y_o = \text{constant} \tag{4b}$$

$$t - \frac{\partial S}{\partial \omega} = t_o = \text{constant} \tag{4c}$$

holds. For constant k_x, k_y, ω, x_o, y_o, t_o, Eqs. (4) determine a curve, or *ray*. Its shape is given by Eqs. (4a) and (4b), and the time at which it reaches a depth z is determined by Eq. (4c). Within the limits of the W-K-B approximation, the rays are the streamlines of the energy flow, and the group velocity is the velocity of the energy. Consequently, energy moves along a ray with group velocity. The shape of the rays is independent of the wavelength and is determined solely by the frequency (Eckart, 1960, pp. 158-162).

Under certain conditions, rays are also the *characteristics* of a partial differential equation in S (identical to the eikonal equation of geometrical optics), so that the terms "ray" and "characteristic" are often used interchangeably.

Magaard (1962), has presented solutions for a standing internal wave in water of variable depth. If the bottom slope is less than the slope of the characteristics, (a) the amplitude of a wave depends on water depth and slope; (b) the amplitude increases with decreasing depth; and (c) the wavelength is dependent only on the depth of the water. The horizontal velocities on the surface and at the bottom, as well as the position of the nodes for u and w, i.e., the depth at which u = 0 and w = 0, are shown in Fig. 4-2. A discussion of Magaard's work may also be found in Krauss (1966b, pp. 58-67).

Other authors, notably Wunsch (1968, 1969), have considered internal waves propagating up a linear slope. The geometry of the

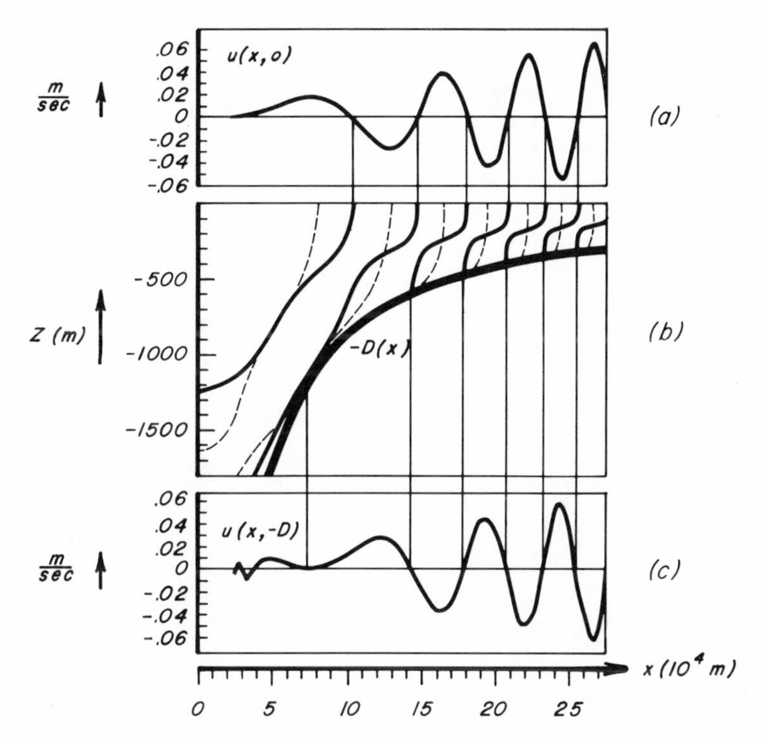

FIG. 4-2. Current distribution of the horizontal, u(x,z), and vertical, w(x,z), velocities for a standing first-mode internal wave in water of variable depth -D(x). (a) u(x,0); (b) depth at which u(x,z) = 0 (solid lines) and at which w(x,z) = 0 (dotted lines); (c) u(x,-D). (Redrawn from Magaard, 1962; used with the author's permission.)

system is shown in Fig. 4-3. The angles of the corners are given by θ_1 and θ_2. If we assume periodic motion, so that the streamfunction is

$$\psi(x,z,t) = \hat{\psi}(x,z)\, e^{-i\omega t}$$

then, neglecting the effects of the earths' rotation, Eq. (3.2.25) becomes

$$\frac{\partial^2\hat{\psi}}{\partial z^2} - \frac{1}{\chi^2}\frac{\partial^2\hat{\psi}}{\partial x^2} = 0 \qquad (5)$$

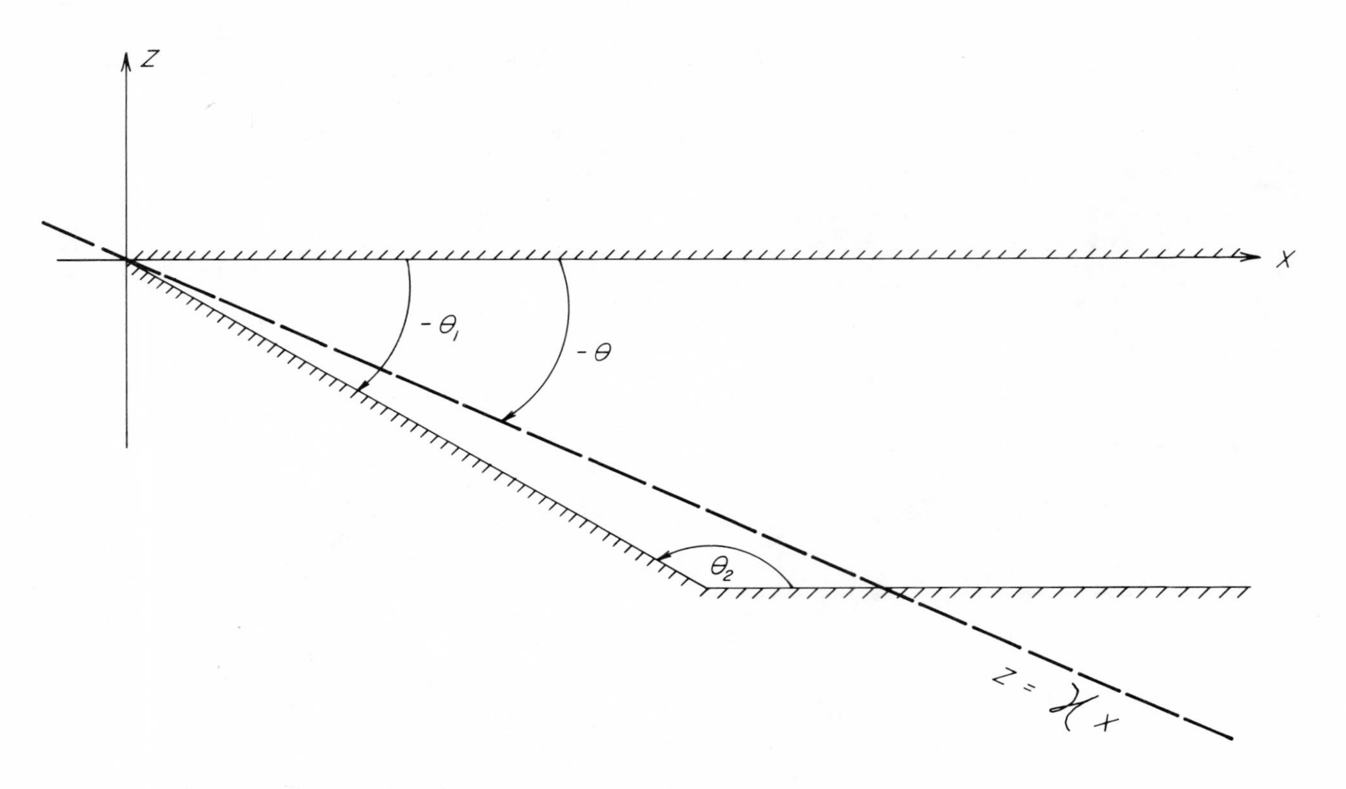

FIG. 4-3. Geometrical description of a wedge. θ_1 and θ_2 are the angles of corners; $-\theta$ is the angle which an internal wave characteristic makes with the horizontal.

where

$$\chi^2 \equiv \frac{\omega^2}{N^2-\omega^2} \tag{6}$$

N is taken to be constant. Wunsch (1968, 1969) had considered solutions to Eq. (5) within a wedge; those solutions which he has found vanish at the top ($\theta' = 0$) and along the bottom slope ($\theta' = -\theta_1$). The solutions have a (removable) singularity along the internal wave characteristic, $\theta' = -\theta$, where $-\theta$ is the angle which the group velocity makes with the horizontal (Secs. 1.7 and 3.4). For sufficiently short waves, the slope of the characteristic is χ. To see this, we use Eq. (3.4.25):

$$\omega = \pm N \sin(-\theta) \tag{7}$$

which holds for sufficiently short waves. Then

$$\chi^2 = \frac{N^2 \sin^2\theta}{N^2 (1 - \sin^2\theta)} = \tan^2\theta$$

so that $\chi = \tan(-\theta)$ holds. [Wunsch denotes the number which we have called "χ" by "c". Equation (8) in Wunsch (1968) contains a misprint (Wunsch 1972: personal communication) and should read $\alpha = \tan^{-1}c$.] Thus (in the fourth quadrant), χ is the slope of the characteristic. Wunsch (1969) has found that if the bottom slope is less than χ, internal waves will propagate upslope, while if the bottom slope is greater than χ, the waves are reflected. Figure 4-3 is drawn for the second case. It appears that there is marked intensification of shear near the bottom slope (cf. Fig. 1-12), and that this shear increases as the slope approaches χ. In addition, the wavelength decreases as the wave shoals. Mooers (1972) has also considered internal and inertial waves propagating up a slope, and solutions near the corner have been considered by Robinson (1970) for restricted values of θ_2.

In a recent paper, Hurley (1972) has discussed a general method for solving steady-state internal wave problems which involve boundaries. The method involves finding solutions for the case $\omega > N$ which are analytic functions of $\omega + i(\delta\omega)$ when $\delta\omega$ is small, and then using analytic continuation to obtain solutions for the case $\omega < N$.

Hall and Pao (1971) have studied the actual breaking process in detail for different beach angles and incident wave conditions (Figs. 1-13, 14, and 15). Their measurements show that the initial breakdown mechanism is a form of shear instability induced by the shearing motion associated with the incident wave.

Among the various aspects of internal motion considered by Hollan (1966a) is the variability of the periods of internal waves caused by variations in depth. Magaard (1968) has looked at free internal waves in a meridional section with variable water depth, as one aspect of considering internal waves as perturbations of geostrophic currents. Niiler (1968) has considered long internal waves in a rotating channel of non-uniform depth. He has shown that the amplitude of an internal seiche (which appears to be caused by the surface tide) is influenced by the topography of the channel. The effect of internal waves on the distribution of sediment on the ocean floor in relation to topography has been considered by Tareev (1965). Dore (1970) has developed nonlinear equations of motion in investigating the effects of finite amplitude on three-dimensional internal waves in a fluid with variable depth. He has found that the second-order first harmonic induced by bottom topography is 90° out of phase with the primary wave.

The reflection of internal or inertial waves from infinite plane surfaces was first discussed in detail by Phillips (1963), and the results for an inviscid fluid have been summarized by Greenspan (1968). Briefly, a plane wave incident on such a surface is reflected as another plane wave whose crests make the same angle (but opposite in sign) to the vertical as those of the incident wave. The reflected wavenumber and energy flux are modified accordingly. Baines (1971a) has considered what happens when plane waves are incident on a smooth surface which is not flat (see Fig. 4-4a) for the case when the incident wavelength is large in comparison with the

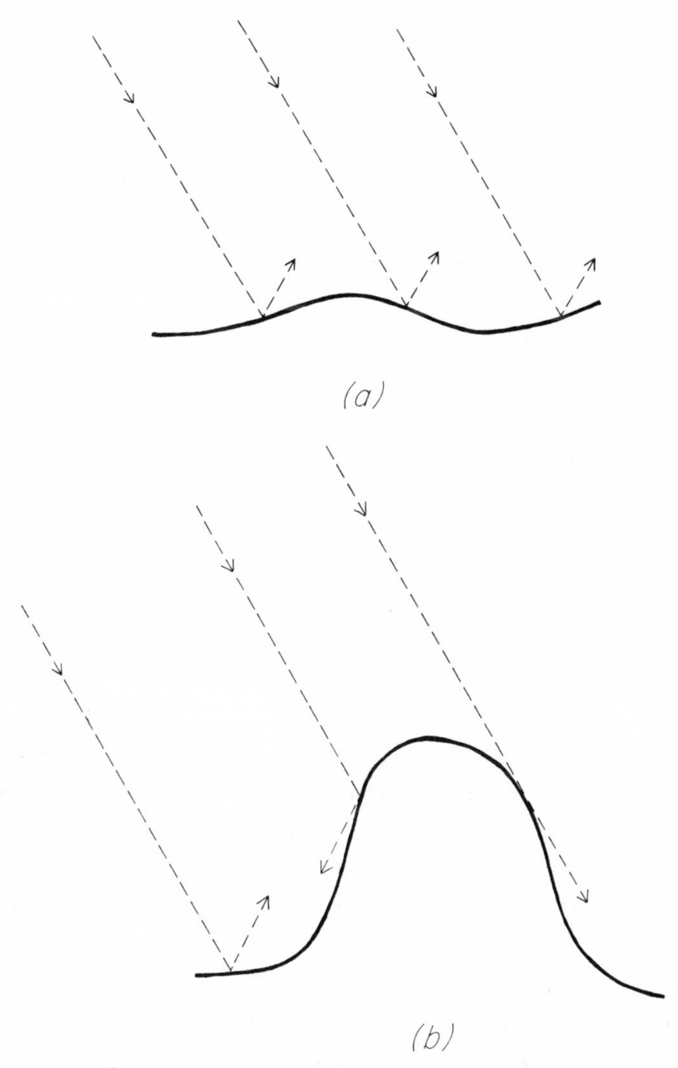

FIG. 4-4. (a) "Flat bump" topography, with rays of the incident wave and their reflections shown; and (b) "steep bump" topography. (Redrawn from Baines, 1971a.)

length scales of the reflecting surface. There is in general some energy flux which is back-reflected in the opposite direction to that of the incident energy flux (in addition to that transmitted in the direction of the reflected rays). Baines has found that the total wave motion is governed by a Fredholm integral equation whose kernel depends on the form of the reflecting surface. If the bottom topography is sinusoidally varying, two new waves, in addition to the main reflected wave, are generated whose wavenumbers are the sum and difference, respectively, of those of the bottom topography and of the incident wave. The sum wave is always in the transmitted direction, whereas the difference wave is back-reflected, if the incident wavelength is sufficiently long. Solutions for general linearized bumps may be obtained by Fourier superposition. Longuet-Higgins (1969) has discussed reflection of internal waves from rough surfaces, but his results should be used with caution (Baines, 1971a). Nagashima (1971) has discussed reflection of internal waves from a sloping beach for a two-fluid system. The reflection of internal waves by a very thin barrier has been considered by Larsen (1969b), Sandstrom (1969), and Robinson (1969).

The refraction of internal waves due to sloping bottoms has been considered by Wunsch (1969), and explicit expressions for the changing wavelengths and amplitudes of two-dimensional waves are given. Wunsch's analysis for three-dimensional waves over small slopes indicates that wave fronts which approach the continental slope at an angle, end up propagating parallel to the slope because of refraction (cf. Fig. 1-4). Dore (1971) has found that bottom friction acts to increase wave refraction slightly.

Baines (1971b) has considered the reflection of internal or inertial waves from bumpy surfaces when a ray is tangent to the surface at some point (Fig. 4-4b). If the ray is associated with the incident wave, diffraction will occur. If the ray is associated with a reflected wave, split reflection occurs, a phenomenon discussed by Baines. In both of these cases, Baines has reduced the problem of determining the wave field to a set of coupled integral equations, with two unknown functions, which are solved for simple topography, and the properties of the wave fields for more general topographies are discussed. For the cases which Baines has considered, instability and subsequent mixing of the fluid is plausible if the incident energy flux is sufficiently large. This criterion is not very demanding, and Baines has suggested that the phenomenon could well have geophysical applications.

Hurley (1970) has discussed diffraction of internal waves around a corner, or wedge (Fig. 4-3) for varying wedge angles. Diffracted waves are as important as the incident and reflected ones at all points that lie within a quarter-wavelength or so of either ray that passes through the apex. Thus, the so-called ray theory for internal waves in which the incident and reflected waves alone are considered has limitations similar to the geometrical theory of optics: both theories involve the assumption that the typical obstacle dimensions in the problem are large compared to the wavelength.

Wave diffraction due to a step change in bottom topography for a two-fluid system has been considered by Kelly (1969); and Manton, Mysak, and McGorman (1970) have considered diffraction of internal waves by a very thin, semi-infinite barrier.

4.5 OTHER CAUSES OF INSTABILITY

Orlanski and Bryan (1969) have proposed that wave motion can produce a horizontal advection of density, causing locally unstable density gradients; that is, $\partial\rho/\partial z > 0$. A criterion for the critical amplitude of a wave is given which they suggest may be more significant than the condition that $Ri < 1/4$.

Penetrative convection occurs when turbulent fluid advances into a fluid layer of stable stratification. In an experiment by Townsend (1964), a tank of water, initially isothermal, was cooled from below and heated gently from above. The lower boundary become cooler than the temperature of maximum density, 4°C, so that an unstable layer developed above it, and at the end of the experiment steady-state conditions were approached. The experiment disclosed several interesting facts. Except in the lowest few centimeters, the temperature fluctuations were of maximum amplitude at or slightly above the level of the mean interface. In the upper stable region these fluctuations appeared to be associated with internal waves excited by the impacts from below of penetrating and subsequently subsiding columns. Fluctuation amplitudes decreased rapidly with further increase in height. In the experimental tank, the length scales were so small that viscous forces dissipated most of the wave energy, but Townsend has surmised that waves in a large tank would lose energy almost entirely by a form of breaking.

Deardorff, Willis, and Lilly (1969), to whom the author is indebted for much of the above summary of Townsend's work, have also done laboratory experiments of nonsteady penetrative convection in water which closely simulate the lifting of an atmospheric inversion above heated ground. In contrast to Townsend's experiment, their mean thermal structure was nonsteady, and the convective layer continually deepened. The interfacial shape appeared to be a combination of forms: domes with adjacent cusps, flat sections, folded structures, and breaking waves (Fig. 1-18).

LeBlond (1966, p. 134) has mentioned a process by which internal waves might become unstable in an ocean where cold, relatively fresh water overlies warmer, more saline water.

4.6 THE CRITICAL LAYER IN THE ATMOSPHERE

In shear flow, a *critical level* for a particular wave $(\omega,\vec{k})$ in the atmosphere is the height z_c where the mean velocity $U_x(z_c)$ [$= U(z_c)$, since $V = W \equiv 0$] of the wind equals the horizontal phase velocity $\omega/(k_x^2+k_y^2)^{1/2}$ of the wave. A *critical layer* for a wave packet containing waves with various frequencies and wavenumbers is a region made up of critical levels for each of the individual waves. Although the only studies on the critical layer to date have been done for atmospheric applications, Booker and Bretherton (1967) have suggested that if there is a vertical transfer of horizontal mean momentum into the deep ocean by the process outlined in this section, it could have "considerable implications in the theory of the general circulation of the oceans."

The following phenomenon has been fully described in an excellent paper by Booker and Bretherton (1967). Suppose first that the flow is stable enough so that the Richardson number Ri(z) is everywhere greater than one, and that the Brunt-Väisälä frequency N is constant. The problem of waves propagating with a nearly horizontal group velocity in the atmosphere is well understood (Bretherton 1972: personal communication), so we confine our attention to packets of waves whose group velocity is, by and large, vertical. Suppose that most of the waves in the packet have frequency ω and wavenumber $\vec{k} = \hat{i}k_x+\hat{j}k_y+\hat{k}k_z$, and suppose that the vertical wavelength λ_z is small (so that the corresponding vertical wavenumber k_z is large). As the packet approaches the critical level where $\omega/(k_x^2+k_y^2)^{1/2} = U(z_c)$, the wave crests tilt forward indicating the upward component of the group velocity (Hazel, 1967, p. 783), the wave amplitudes increase, and the frequency ω decreases relative to the surrounding fluid. This is associated with a decrease in vertical wavelength, small to begin with, and the group velocity becomes more and more horizontal. When the wave packet gets close enough to the critical level, the waves are absorbed and the surplus momentum is taken up by the mean flow in a thin layer just below the critical level. This effect has been demonstrated in the laboratory (Appendix to Hazel, 1967). Thus, for all practical purposes, the waves are confined to the region below the critical level. Figure 4-5 depicts schematically what happens. The over-all result is that after a long time the waves in a wave packet will redistribute themselves below their respective critical levels.

If the Richardson number is between 1/4 and 1, the waves may pass through the critical level, but their amplitudes will be attenuated by a factor $\exp\left\{-2\pi[Ri(z_c) - 1/4]^{1/2}\right\}$. Thus, for $Ri > 1/4$, the critical level effectively acts as an absorber. The mechanism is independent of viscosity and heat conduction, although these may alter the waveform (Hazel, 1967).

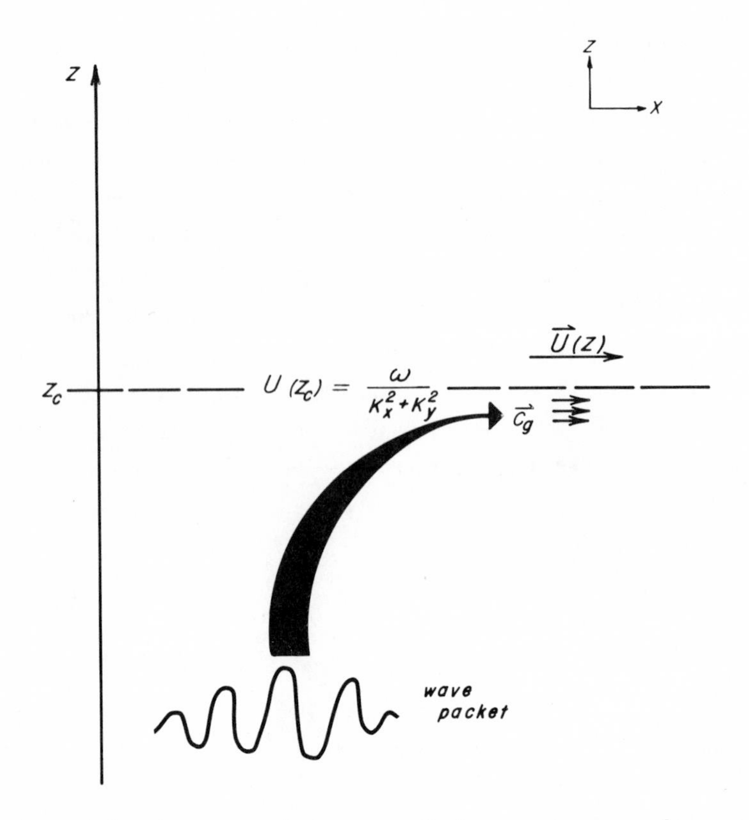

FIG. 4-5. A wave packet, most of whose waves have frequence ω and wavenumber $\vec{k} = \hat{i}k + \hat{j}k_y + \hat{k}k_z$, approaching a critical layer at group velocity $\vec{c}_g$.

Jones (1968) has considered the above process for unstable, flows for which $0 < Ri < 1/4$, and has found numerically that a substantial amount of reflection occurs. In fact, for any given wave there is a critical value for Ri below which the reflected wave is larger than the incident wave. This condition stems from the ability of the wave to extract energy and momentum from the mean flow.

Kelly and Maslowe (1970) have considered a nonlinear critical layer, with possible implications for CAT.

4.7 INTERNAL WAVES AND OCEAN CURRENTS

While the interaction of internal waves and currents is potentially of great importance to oceanographers because of the possibilities of energy transport and mixing, the problem seems to have received little attention thus far. The last section dealt with one aspect of internal wave-current interaction. In this section we first consider the effects of ocean currents on internal waves in a two-fluid system, then in a continuously stratified system, and finally look at a few effects of internal waves upon ocean currents.

In a two-fluid system, Haurwitz (1948) has investigated internal waves in which both layers have a mean velocity. Iwata (1962) has considered only a mean velocity in the lower layer, while Wang (1969) has allowed the mean current to vary horizontally, as well as vertically. A general result is that if the waves propagate with the current, their wavelengths increase and their amplitudes decrease; if they propagate against the current, their wavelengths decrease and their amplitudes increase. In reality, this latter effect may be less than predicted, according to wavetank experiments by Phillips, George, and Mied (1968).

The following results hold if N is constant and if there is a uniform current with a weak vertical shear (i.e., if the Richardson number is much greater than 1/4):

1. Lines of constant phase are rotated toward the vertical. With geostrophic flow, the lines of constant phase tend to wrap themselves around the core of the flow and to converge to the center of the flow (Mooers, 1970).
2. The wavelength decreases, and energy is transferred toward the shorter wavelengths.
3. The group velocity decreases with time, and the wave loses energy. Energy loss from the waves is compensated by a small energy gain in the mean current.
4. The frequency is lowered.

5. The vertical velocity dies most rapidly, and the motion becomes almost entirely horizontal. [This idea is used by Frankignoul (1970a) to explain the generation of inertial waves.]
6. As the vertical velocity decreases, the vertical gradient of the velocity continues to increase and turbulence may develop.

The above results have been discussed by Phillips (1966, pp. 178-185) and by Frankignoul (1970b). Results (2) and (3) have been shown to hold even when a general spatial variation in the mean flow is permitted (Jones, 1969a). The frequency and energy of internal waves propagating in a shear flow have been investigated by Bretherton (1969). Hollan (1966a) has considered the influence of a mean shear current on the wavelength of short internal waves in a shallow sea.

A uniform current has the effect of shifting the frequency limits for internal waves away from their theoretical values of f and N (Saint-Guily, 1970; Magaard, 1968). If the waves propagate with the current, the limits are raised; if they propagate against the current, the limits are lowered. A rigorous treatment of the effect of sheared currents on the limit N and on other properties of internal wave modes is given by Bell (1974a).

For waves propagating in a uniform current $\vec{U}$, the relationship

$$\frac{d}{dt}\frac{E}{\omega'} + \frac{E}{\omega'}(\nabla \cdot \vec{c}_g) = 0$$

holds, where $\omega' = \omega - (\vec{U}\cdot\vec{k})$ is the frequency which would be measured by an observer moving with the local mean velocity of the medium, and E is the wave energy density (Bretherton and Garrett, 1968).

The interaction between internal waves and a nonsteady, nonuniform current has been considered by Garrett (1968) and Wang (1969). If S_{ij} is a tensor describing the stress due to waves and called the *radiation stress tensor* (Phillips, 1966, p. 170) and if T_{ij} is a tensor describing the stress which works against the rate of strain of the basic flow and called the *interaction stress tensor*, then for the case of a current described by

$$\vec{u} = \hat{i}u_x(x,y,z,t) + \hat{j}v(x,y,z,t) + \hat{k}w(z,t)$$

ERRATA

ROBERTS: *Internal Gravity Waves in the Ocean*

Page 187 – lines 10 through 14 have been inadvertently positioned in the wrong place. They should appear at the bottom of the page.

we have

$$T_{33} = S_{33}$$

(Garrett, 1968). Bretherton (1969) and Chunchuzov (1971) have considered the upward transport of horizontal momentum by internal waves in the atmosphere and the effect on the mean flow. Wang (1969) also has shown that internal waves and the current can exchange momentum and energy. In fact, theoretical evidence is mounting for the existence of significant interactions between inertial and internal waves, and geostrophic flows (Moors, 1970).
it will be either reflected or dissipated (Phillips, 1966, p. 196). Critical latitudes for the internal tides are of particular interest (Knauss, 1962; Hendershott, 1968). Magaard (1968) has shown that ocean currents may bring about considerable displacement of the critical latitudes.

For internal waves of finite amplitude, Frankignoul (1972) has found that wave theory may be inadequate for describing energy exchanges between the waves and a simple shear flow. Breeding (1972) and Grimshaw (1971) have also treated the problem for nonlinear waves. One effect of the nonlinear terms is ultimately to alter the vertical structure of a shear flow (Grimshaw, 1971). Breaking internal waves may even generate a longshore current (Hogg, 1971).

4.8 CRITICAL LATITUDES

The period T of a free internal gravity wave is constrained to lie within the limits $T_N < T < T_f$, where T_N is the Brunt-Väisälä period and T_f is the inertial period, which decreases from infinity at the equator to about 12 hr at the poles (Sec. 3.3 discusses the limits for T). For a particular wave of period T_o, the latitude for which $T_o = T_f$ is called a *critical latitude* for that wave. When an internal wave, propagating poleward, encounters its critical latitude,

4.9 DAMPING

The following are some facts concerning the damping of internal waves:

1. Internal waves are much more strongly damped than are surface waves (Crampin and Dore, 1970; Fedosenko, 1969).
2. In a two-fluid system, viscosity decreases the wavelength of both short (Harrison, 1908) and long (Rattray, 1957; Walker, 1972) internal waves.
3. In a two-fluid system, internal waves are damped more when the interface is near the surface or near the bottom (Thorpe, 1968c; Heaps and Ramsbottom, 1966).
4. Higher-mode internal waves are more heavily damped than those of lower mode numbers (Crampin and Dore, 1970; Johns and Cross, 1970).
5. Internal waves are most strongly damped in a layered system, least damped in a system which has a sharp thermocline (Crampin and Dore, 1970).
6. Thermal diffusion may be as important as viscosity in damping internal waves in most of the world oceans (LeBlond, 1970).
7. In an exponentially stratified ocean, very short internal waves with $\lambda < 100$ m are strongly damped (LeBlond, 1966).
8. In an exponentially stratified ocean, long internal waves in very shallow water ($D = 100$ m) are strongly damped, while in deeper water ($D = 4\times10^3$ m) they are less damped (LeBlond, 1966).
9. Friction decreases the amplitudes of long internal waves over a shelf, changing them from standing waves to ones that progress shoreward (Weigand *et al.*, 1969).
10. The extent to which the amplitudes of shoaling internal waves are damped depends upon the angle of incidence and on the ratio

    $$(R_w)^{-1/2} : \tan(-\theta_1)$$

    where the wave Reynolds number R_w is defined as

    $$R_w \equiv \frac{2\,\omega}{\bar{\nu} k^2}$$

    $\bar{\nu}$ is a representative value for the kinematic viscosity, and $\tan(-\theta_1)$ is the slope of the beach (Fig. 4-3). A moderately turbulent boundary layer can have appreciable effect on the wave amplitude (Dore, 1971).

Crampin and Dore (1970) have done a numerical comparison of damping of internal waves in fluids with different density profiles: (a) homogeneous fluid (H); (b) thermocline density profile (T); (c) exponential profile (E); (d) two-fluid, or discontinuous, system (D). For the different profiles, they give x_τ/λ, the *damping length*, as a function of wavenumber (Fig. 4-6). Here x_τ is the distance traveled by a wave during the time τ, and λ is its wavelength. A few results may be noted from Fig. 4-6. First-mode waves are damped less than second-mode waves, and short waves are damped more than long ones. Relative damping lengths in turbulent flow are shown in Fig. 4-7. This figure represents qualitative effects of

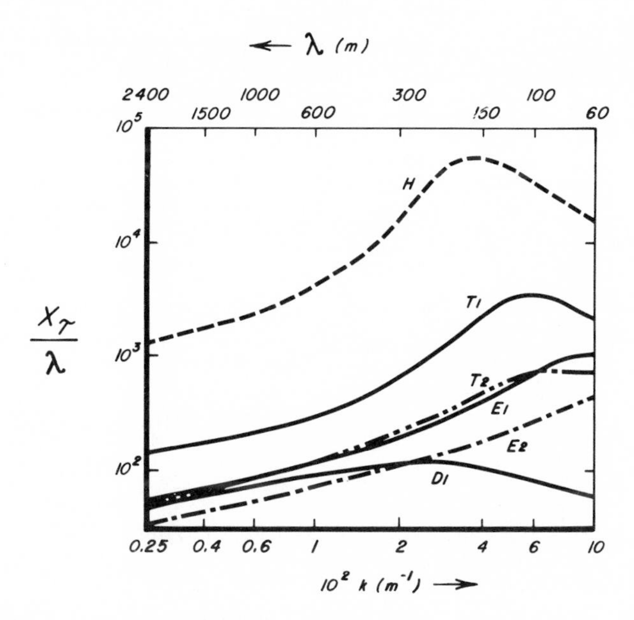

FIG. 4-6. Relative damping lengths in laminar flow. Letters refer to the particular density profile which was assumed for the computation, subscripts indicate internal wave mode numbers. The water depth is -100 m. (Redrawn from Crampin and Dore, 1970; used with Dr. Dore's permission.)

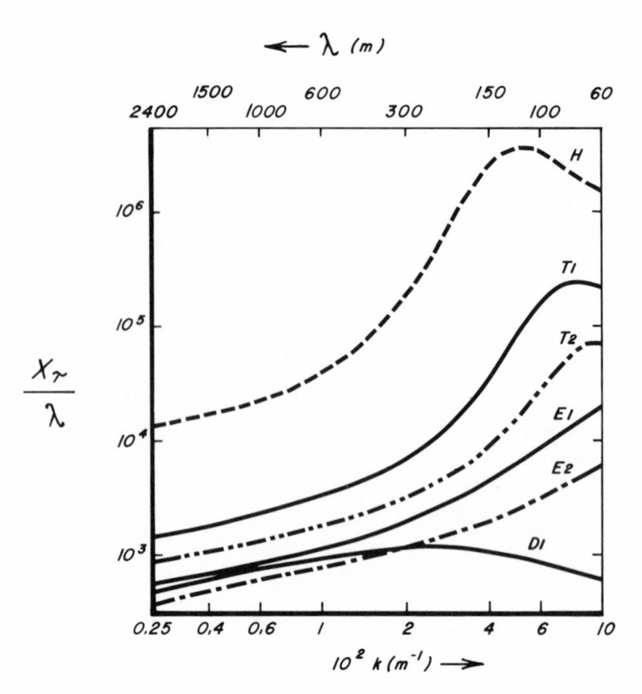

FIG. 4-7. Relative damping lengths in turbulent flow. The water depth is -100 m. (Redrawn from Crampin and Dore, 1970; used with Dr. Dore's permission.)

turbulence, obtained by assuming a constant coefficient of eddy viscosity for momentum diffusion. For each density distribution, Crampin and Dore have selected a value for the coefficient of 10^2 times the molecular value.

Terminal amplitudes due to viscosity in the presence of resonant wave interactions have been given by Davis and Acrivos (1967b), McEwan (1971), and Orlanski (1972). Viscous effects at the walls of wavetanks have been measured by Keunecke (1970) and calculated by Kelly (1970) and Thorpe (1968c). Changes in amplitude and phase in a viscous fluid have been discussed by Thomas and Stevenson (1972), Gordon and Stevenson (1972), and Grimshaw (1973). Grimshaw (1973) has also considered the mean flow of an internal wave packet in a viscous, slightly compressible fluid.

4.10 SURFACE EFFECTS

That internal waves moving through shallow waters may be associated with bands of slicks on the surface has been well documented by LaFond (for instance, LaFond, 1966) and Ewing (1950). The usual explanation for slicks is that the alternate convergence and divergence of the surface current due to internal waves will cause a periodic contraction and expansion of the surface film (which is composed of organic debris), with maximum extension of the film over internal wave crests and maximum compression over the troughs. When the film is compressed it will damp out surface waves much more rapidly than when it is extended and hence, the argument goes, the troughs of internal waves will be indicated on the surface by smooth bands, or slicks, while the crests will be marked by bands of rough water. As mentioned in Sec. 1.9, internal waves whose crests are close enough to the surface may even cause breakers to appear on the surface.

Gargett and Hughes (1972) have suggested, however, that the mechanism for the actual formation of these bands may at times be more complicated than originally supposed. Their study was prompted by the observation that surface waves seemed to change their direction of propagation over the crests of internal waves (Fig. 4-8). They therefore investigated the interaction between surface waves and a periodic mean current induced by an internal wave. Some of their results are as follows: If $c_{g_s} (\cos \phi) > c_{p_i}$ over an internal wave crest, that is, if the surface wave groups overtake the internal wave, then the surface waves will tend to propagate in the direction of the internal wave and their amplitudes will increase. If $c_{g_s} (\cos \phi) < c_{p_i}$ over an internal wave crest, that is, if the surface wave groups are overtaken by the internal wave, the direction of the surface waves is turned away and their amplitudes decrease. Over troughs, the opposite effects occur. Here c_{g_s} is the surface wave group speed, c_{p_i} is the internal wave phase speed, and ϕ is the angle which $\vec{c}_{g_s}$ makes with $\vec{c}_{p_i}$. Their analysis predicts that for a

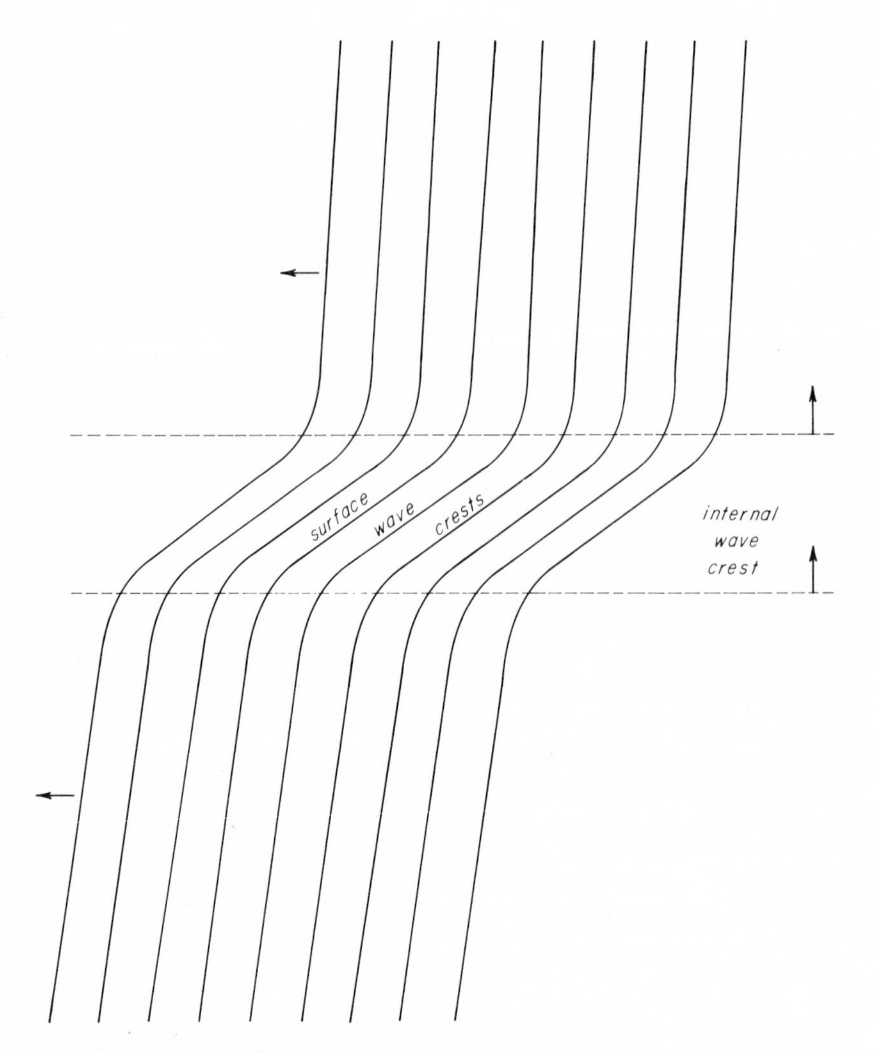

FIG. 4-8. Surface wave crests changing direction as they propagate over an internal wave crest (Hughes 1972: personal communication).

given internal wave field, the system will have singularities for certain surface waves. Therefore, there will be regions where none of these surface waves can exist, and these regions will appear as surface slicks.

While the surface currents associated with internal waves may cause noticeable effects, the vertical displacement of the surface due to internal waves is slight and is usually neglected. The internal tide may, however, cause a measurable surface displacement. Radok, Munk and Isaacs (1967) have found that the internal tide in the open ocean may account for an incoherent 1-cm contribution to the elevation of the surface tide. Miles (1971) has given an expression for the surface displacement caused by internal waves generated by a moving body.

4.11 MIXING

Breaking internal waves may be an important mechanism for oceanic mixing. Rooth (1971) has considered oceanic mixing by internal waves breaking due to resonant instability; Garrett and Munk (1972a) have asserted that mixing is much more likely to be caused by shear instability than by direct overturning of the waves. Glinskii and Boguslavskii (1963) and Boris (1970b) have investigated mixing due to internal wave-caused turbulence without specifying how the waves broke. Glinskii and Boguslavskii have shown that internal waves with shorter periods (around 7 hr) are much more important in vertical exchange than internal waves of longer period (around 68 hr). Using the internal wave amplitude a, as determined numerically from hydrographic data, Boris (1970b) has constructed maps showing seasonal variation in the North Atlantic of a maximum kinematic turbulent viscosity coefficient A_ν,

$$A_\nu \equiv \frac{4\pi\chi^2 a^2}{18T}$$

where χ is von Kármán's constant (Bird *et al.*, 1960, p. 160) and T is the internal wave period, taken to be 12.5 hr. She has shown that A_ν has the highest values in June off Cape Blanc, Africa, with the next highest values, also in June, in the Sargasso Sea; the lowest values for A_ν occur in January, February, and August. Maps showing seasonal variability of the vertical current (the vertical velocity associated with internal waves, according to Boris, is 10^2 higher than the vertical velocity for other types of currents) follow the same pattern, leading her to conclude that "internal tidal waves have a significant effect on the vertical water exchange in large regions of the North Atlantic." This, coupled with Garrett and Munk's (1972a) observation that the horizontal eddy diffusivity due to breaking internal waves is about $10^4 N^2$ greater than the vertical eddy diffusivity, leads one to conclude that internal waves may indeed be very important to mixing processes in the world oceans.

4.12 MICROSTRUCTURE

Microstructure, or *fine-structure*, as it is sometimes called, is a term loosely used to describe some departure of the values of temperature, salinity, or density from a mean value. A recent review of oceanic microstructure has been given by Gregg (1973). While it has been known for some time that stratified water tends to layer (U. of Calif. Div. of War Research, 1942), the primary mechanisms for the formation of oceanic microstructure are still a subject for some conjecture (for instance, see Phillips, 1972). It has been attributed variously to internal waves breaking by shear instability (Garrett and Munk, 1972a); to internal waves breaking as they shoal, or instability of boundary layers on a slope (Wunsch and Dahlen, 1970b); to long-period internal waves (Lazier, 1973); to salt fingering (for example, Stern and Turner, 1969); to viscous overturning (Baker, 1972); to sporadic overturning associated with finite amplitude internal waves (Orlanski and Bryan, 1969); and to turbulent flow (Phillips, 1972).

Whatever its origin, the presence of microstructure can considerably complicate the interpretation of internal wave data (Phillips, 1971; Garrett and Munk, 1971; Joyce, 1973; Munk and Garrett, 1973), as well as reduce the coherence of the vertical velocity associated with internal waves (Miropol'skii, 1972). It also appears that the high density gradients associated with microstructure provided a preferential site for shear instability of internal waves (Garrett and Munk, 1972a).

POSTSCRIPT

Many papers on internal waves have appeared since I completed this manuscript in 1973. Some of these are listed in References Added in Press. The size of that section, which lists publications since 1972, compared to that of the Bibliography, which lists publications as early as 1889, gives some idea of the large amount of work which has been done on internal waves in the last two years. Briscoe (1975, References Added in Press) has written a survey of this work.

The last two years have also seen a shift from a description of internal waves as analytic solutions to the vertical velocity equation, as is done in Chap. 3, to a description of internal waves in terms of their spectra. This shift in emphasis is largely due to the paper by Garrett and Munk (1972b), in which they proposed a wave number-frequency spectrum, based on then-available observations. In that paper they invited and encouraged more oceanic observations of internal waves, and they have recently published a revised spectral model using these newer observations (Garrett and Munk, 1975, References Added in Press).

I had originally included a chapter titled "Areas for Investigation," but the results of the past two years have rendered many of its suggestions obsolete. Generation and interaction problems are much better understood now (Bell, 1975, and Olbers, 1974, respectively; References Added in Press), and Garrett and Munk have provided oceanic internal wave observers with a much-needed sense of direction. Thorpe (1975, References Added in Press) and Wunsch (1975, References Added in Press) have indicated areas of internal wave study which currently need more work.

Fairbanks, Alaska
May, 1975

NOTATION

Unless otherwise stated, numbers in parentheses after description refer to the section in which the symbols are first used or thoroughly defined.

A	Constant
A_ν	Maximum kinematic turbulent viscosity coefficient (4.11)
a	Wave amplitude
a	Acceleration
B, B_1, B_2	Constants
C_0, C_1, etc.	Constants
$\vec{c}_g$	Group velocity of a wave train (3.4); $\vec{c}_g = \hat{i}c_{g_x} + \hat{j}c_{g_y} + \hat{k}c_{g_z}$
$\vec{c}_{g_s}$	Group velocity of a train of surface waves (4.10)
c_i	Imaginary part of the phase speed $c_p = \lvert\vec{c}_p\rvert$ (3.9)
$\vec{c}_p$	Phase velocity of a wave; $\vec{c}_p = \hat{i}c_{p_x} + \hat{j}c_{p_y} + \hat{k}c_{p_z}$
$\vec{c}_{p_i}$	Phase velocity of an internal wave (4.10)
c_r	Real part of the phase speed $c_p = \lvert\vec{c}_p\rvert$ (3.9)
cn x	Jacobian elliptic function (3.11)
$-D$	Total water depth; usually constant, except in Sec. 4.4
$-d$	Depth of the thermocline; it may be an average depth, or it may be the depth of discontinuity of a two-fluid system

d_1	d (3.4)
d_2	D - d (3.4)
E	Wave energy density (3.6)
E_{pot}	Potential energy (3.6)
E_{tot}	Total energy density per unit mass (3.6)
exp x	Exponential function of x; e^x
F(x,y)	amplitude of the vertical velocity w (3.4); F is a function of x and y only
F(A,B,x)	Hypergeometric function (3.4)
$\vec{F}_b$	Buoyancy force (3.4)
F_{tot}	Total energy flux (3.6)
$\vec{\mathcal{F}}$	Arbitrary force; $\vec{\mathcal{F}} = \hat{i}\mathcal{F}_x+\hat{j}\mathcal{F}_y+\hat{k}\mathcal{F}_z$
$\mathcal{F}(t)$	$\lvert\vec{\mathcal{F}}(t)\rvert$
$\tilde{\mathcal{F}}(\omega)$	Fourier transform of $\mathcal{F}(t)$
f	Twice the vertical component of the earth's angular velocity (3.2); $f = 2\Omega_z$
f_1, f_2	Driving forces responsible for second-order internal waves (2.2)
G(x)	Amplitude of the horizontal velocity u_x (3.4); G is a function of x only
$G(\omega,\vec{k})$	Dispersion relation (2.2)
$\mathcal{G}(t)$	Green's function (2.9)
$\vec{g}$	Gravitational acceleration
h	Vertical length scale (5.4)
h_1	d (3.11)
h_2	D - d (3.11)
i	$\sqrt{-1}$
$\hat{i}$	Unit vector in the x-direction

$J(\Delta\phi',\phi')$ — Jacobian (3.5); since $\Delta\phi'$ and ϕ' are functions of s and z, $J(\Delta\phi',\phi') = \partial(\Delta\phi',\phi')/\partial(s,z)$

$J_m(r)$ — Bessel function of the first kind of order m (3.4)

$\mathcal{J}$ — Defined by Eq. (3.4.41)

$\hat{j}$ — Unit vector in the y-direction

K — Parameter for Whittaker function (3.4)

$\hat{k}$ — Unit vector in the z-direction

$\vec{k}$ — Wavenumber vector; $\vec{k} = \hat{i}k_x+\hat{j}k_y+\hat{k}k_z$

$\vec{k}_h$ — Horizontal wavenumber vector; $\vec{k}_h = \hat{i}k_x+\hat{j}k_y$

k_h — Horizontal wavenumber; $k_h^2 = |\vec{k}_h|^2 = k_x^2 + k_y^2$

k_x — Wavenumber in x-direction; $k_x = 2\pi/\lambda_x$

k_y — Wavenumber in y-direction; $k_y = 2\pi/\lambda_y$

k_z — Wavenumber in z-direction; $k_z = 2\pi/\lambda_z$

ℓ — (a) width of continental shelf (2.4); (b) layer thickness (3.4)

ln x — logarithm of x to the base e

$M_{k,\sigma/2}$ — Whittaker function (3.4)

m — Order of Bessel function (3.4)

N — Brunt-Väisälä frequency (3.3); $N^2 = (-g/\bar{\rho})(d\bar{\rho}/dz)$

n — Mode number for an internal wave (3.4)

p* — Pressure (3.2); $p^* = \bar{p}(z) + p(x,y,z,t)$

p — Fluctuations of pressure resulting from fluid motion, a function of x,y,z, and t (3.2)

$\bar{p}$ — Average pressure (3.2); $\bar{p}$ is a function of z only

p_0 — Hydrostatic pressure in an ocean at rest with constant density ρ_0 (Appendix 3.2.1)

$\mathcal{Q}$ — Nonlinear terms in Eq. (3.2.14)

q(z) — $[N^2(z) - \omega^2]\bar{\rho}(z)$ (3.4)

$\dot{q}$	Rate at which heat increases in a unit volume (3.2)
$\vec{R}$	Right-hand side of Eq. (1) in Appendix 3.2.2
$R(\omega)$	Response function (2.3); the Fourier transform of Green's function $G(t)$
Re	Reynolds number (3.9)
Ri	Richardson number (3.9)
R_w	Wave Reynolds number (4.9)
$\vec{r}$	Radius or position vector; $\vec{r} = \hat{i}x + \hat{j}y + \hat{k}z$
r	Parameter for Bessel function only in (3.4)
S	$k_z z$, a function of k_x, k_y, ω, and z (4.4)
S_{ij}	Radiation stress tensor (4.7)
$\mathcal{S}$	Right-hand side of Eq. (7) in Appendix 3.2.2
s	(a) displacement (3.4b); (b) $s = x - c_p t$ (3.5)
T	Wave period; $T = 2\pi/\omega$
T_f	Inertial period (3.3); $T_f = 2\pi/f = \pi/(\Omega \sin\phi)$
T_N	Brunt-Väisälä period (3.3); $T_N = 2\pi/N$
T_{ij}	Interaction stress tensor (4.7)
t	time
$\vec{U}$	time-independent current vector, i.e., the velocity of an ocean current or of the wind (3.2); $\vec{U} = \hat{i}U_x + \hat{j}V + \hat{k}W$
U'	Magnitude of current speed in Eq. (3.9.14)
U	$\lvert\vec{U}\rvert$
U_x	x-component of current, a function of x,y, and z
$\mathcal{U}(z)$	Amplitude of the horizontal velocity u_x (3.4)
$\vec{u}^*$	Instantaneous velocity vector (3.2); $\vec{u}^* = \vec{u} + \vec{U}$; $\vec{u}^* = \hat{i}u_x^* + \hat{j}v^* + \hat{k}w^*$
$\vec{u}$	Time-dependent fluctuations of velocity (3.2); $\vec{u} = \hat{i}u_x + \hat{j}v + \hat{k}w$

u	$\lvert\vec{u}\rvert$
u_x	Fluctuation of velocity in the x-direction
V_s	Speed of sound in a fluid
v	Fluctuation of velocity in the y-direction
W	z-component of current
$\mathcal{W}$	Amplitude of the vertical velocity w (3.4); $\mathcal{W}$ is a function of z only
$\mathcal{W}_n$	Amplitude of the vertical velocity w_n for an nth order internal wave (3.4)
w	Fluctuations of velocity in the z-direction
$\hat{w}$	The time-independent part of w (3.3)
X	Function of x only, used to separate F(x,y) (3.5)
$\mathcal{Y}$	Real part of $\mathcal{W}$ (3.9)
x	Horizontal axis
x_τ	Distance traveled by a wave during time τ (4.9)
Y	Function of y only, used to separate F(x,y) (3.5)
$Y_m(t)$	Bessel function of the second kind of order m (3.4)
	Imaginary part of (3.9)
y	Horizontal axis
$y(t)$	Arbitrary variable (2.3)
$\tilde{y}(\omega)$	Fourier transform of y(t) (2.3)
z	Vertical axis pointing up
z_o, z_1, etc.	Fixed, specified depths
α	Constant
α_o, α_1, α_2	Defined in Sec. (4.2)
β	Constant
Γ	Gamma function (3.4)

γ	Interfacial surface tension (4.3)
δ	Infinitesimal
$\delta(t)$	Unit impulse function (2.3)
ε	Thermocline thickness or density gradients (3.4)
ζ	z-component of fluid displacement caused by a wave
η	Variable defined in Sec. (3.4)
θ	Angle which group velocity makes with the horizontal
θ_1, θ_2	Angles of the corners of a wedge (4.4)
θ'	Arbitrary angle (4.4)
$\vec{\lambda}$	Wavelength vector; $\vec{\lambda} = \hat{i}\lambda_x + \hat{j}\lambda_y + \hat{k}\lambda_z$
μ	Coefficient of viscosity (3.2)
ν_n	n^{th} eigenvalue of the internal wave Eq. (3.4)
ρ^*	Density (3.2); $\rho^* = \bar{\rho}(z) + \rho(x,y,z,t)$
ρ	Fluctuations of density resulting from fluid motion, a function of x, y, z, and t (3.2)
$\bar{\rho}$	Average density, a function of z only
ρ_0	Average density evaluated at the surface; $\rho_0 = \bar{\rho}(0)$
ρ_{-d}	Average density evaluated at depth -d; $\rho_{-d} = \bar{\rho}(-d)$
ρ_n	Fluctuations of density caused by an nth order internal wave (3.6)
σ	Constant
ϕ	(a) latitude; (b) the angle between $\vec{c}_{g_s}$ and $\vec{c}_{p_i}$ (4.10)
ϕ'	Modified streamfunction (3.5); $\phi'(s,z) = \psi(s,z) - c_p z$
χ	(a) slope of an internal wave characteristic (3.4); (b) von Kármán's constant (4.11)
ψ	Streamfunction (3.2); $u = \partial\psi/\partial z$, $w = -\partial\psi/\partial x$
$\hat{\psi}$	Amplitude of the streamfunction (4.4)
$\vec{\Omega}$	Angular velocity of the earth (3.2); $\vec{\Omega} = \hat{i}\Omega_x + \hat{j}\Omega_y + \hat{k}\Omega_z$

ω	Angular frequency of a wave; $\omega = 2\pi/T$
ω'	$\omega - \vec{U} \cdot \vec{k}$ (4.7)

MATHEMATICAL OPERATIONS

$\frac{D}{Dt}$	Substantial derivative operator (3.2); $\frac{D}{Dt} = \frac{\partial}{\partial t} + (u^* \cdot \nabla)$
D	Linear differential operator (2.3)
∇	Nabla, or del, operator; $\nabla = \hat{i}\frac{\partial}{\partial x} + \hat{j}\frac{\partial}{\partial y} + \hat{k}\frac{\partial}{\partial z}$
∇^2	Laplacian operator; $\nabla^2 = \frac{\partial^2}{\partial x^2} + \frac{\partial^2}{\partial y^2} + \frac{\partial^2}{\partial z^2}$
∇_h	Horizontal Nabla operator; $\nabla h = \hat{i}\frac{\partial}{\partial x} + \hat{j}\frac{\partial}{\partial y}$
∇_h^2	Horizontal Laplacian operator; $\nabla_h^2 = \frac{\partial^2}{\partial x^2} + \frac{\partial^2}{\partial y^2}$
*	Denotes complex conjugates (except in Sec. 3.2)

REFERENCES ADDED IN PRESS

The following books and papers have appeared since the manuscript for this book was written. The report by Briscoe (1975) contains additional references, many in Russian. Numbers in parentheses after each reference are section numbers in the text appropriate to the reference.

Ahrnsbrak, W. F. (1974). Some additional light shed on surges. *J. Geophys. Res.* 79(24), 3482-3483 (3.8).

Apel, J. R., H. M. Byrne, J. R. Proni, and R. L. Charnell (1975). Observations of oceanic internal and surface waves from the earth resources technology satellite. *J. Geophys. Res.* 80(6), 865-881 (1.3, 1.4, 1.9).

Arhan, M. (1973). "Etude non Linéaire des Ondes Internes dans un Milieu à Deux Couches Fluides en Rotation." Centre National pour l'Exploitation des Océans, Brest, France, Tech. Rept. 17, 29 pp. (3.10).

Banks, W. H., H. P. G. Drazin, and M. B. Zaturska. On the Normal Modes of Parallel Flow of Inviscid Stratified Fluid. School of Mathematics, Univ. Bristol, England. Unpub. manuscript. (3.9).

Barbee, W. B., J. G. Dworski, J. D. Irish, L. H. Larsen, and M. Rattray, Jr. (1975). Measurement of internal waves of tidal frequency near a continental boundary. *J. Geophys. Res.* 80 (in press) (1.3, 1.5).

Barnett, T. P. and R. L. Bernstein (1975). Horizontal scales of mid-ocean internal tides. *J. Geophys. Res.* 80 (in press) (1.5).

Bell, T. H., Jr. (1974). Internal wave-turbulence interpretation of ocean fine structure. *Geophys. Res. Letters* 1, 253-255 (4.12).

Bell, T. H., Jr. (1974). Processing vertical internal wave spectra. *J. Phys. Oceanogr.* 4, 669-670 (1.10).

Bell, T. H., Jr. (1975). Lee waves in stratified flows with simple harmonic time dependence. *J. Fluid Mech.* 67, 705-722 (3.9).

Bell, T. H., Jr. (1975). Momentum and energy transport by mesoscale processes in the deep ocean. *Mém. Soc. Roy. Sci. Liège* 6 (in press) (2.4, 4.9, 4.11).

Bell, T. H., Jr. (1975). A numerical study of internal wave propagation through ocean fine structure. *Dtsch. Hydrogr. Zeit.* 28 (in press) (3.4, 4.12).

Bell, T. H., Jr. (1975). Topographically generated internal waves in the open ocean. *J. Geophys. Res.* 80(3), 320-327 (2.4).

Bell, T. H., Jr., J. Bergen, J. Dugan, Z. Hamilton, B. Morris, B. Okawa, and E. Rudd. Internal waves: measurements of the two-dimensional spectrum in vertical/horizontal wavenumber space. *Science* (submitted 1975) (1.4, 1.5).

Bell, T. H. Jr., A. B. Mays, and W. P. de Witt (1974). "Upper Ocean Stability, a Compilation of Density and Brunt-Väisälä Frequency Distributions for the Upper 500 m of the World Ocean." Naval Research Lab., Washington, D.C., Vol. 1, "Profiles." NRL Rept. 7799, 442 pp.; Vol. 2, "Tables." NRL Rept. 7800, 442 pp. (3.3, 3.4).

Boyce, F. M. (1974). Some aspects of Great Lakes physics of importance to biological and chemical processes. *J. Fish. Res. Bd., Canada.* 31(5), 689-730 (1.5, 4.11).

Brekhovskikh, L. M., K. V. Konjaev, K. D. Sabinin, and A. N. Serikov (1975). Short period internal waves in the sea. *J. Geophys. Res.* 80(6), 856-864 (1.4).

Briscoe, M. G. (1975). Introduction to collection of papers on oceanic internal waves. *J. Geophys. Res.* 80(3), 289-290 (1.1 3.1).

Briscoe, M. G. (1975). Internal waves in the ocean: 1975. *Rev. Geophys. Space Phys.* 13(3), 591-598.

Briscoe, M. G. (1975). Preliminary results from the tri-moored internal wave experiment (IWEX). *J. Geophys. Res.* 80 (in press) (1.2, 1.4, 1.5).

Cacchione, D. A. and J. B. Southard (1974). Incipient sediment movement by shoaling internal gravity waves. *J. Geophys. Res.* 79(15), 2237-2242 (1.8, 1.9).

Cacchione, D. A. and C. I. Wunsch (1974). Experimental study of internal waves over a slope. *J. Fluid Mech.* 66(2), 223-239, 3 plates (1.8).

Cairns, J. L. (1975). Internal wave measurements from a midwater float. *J. Geophys. Res.* 80(3), 299-306 (1.4, 1.5).

Cavanie, A. G. (1973). Equations des ressants internes dans un système à deux couches et leurs solutions. *Mém. Soc. Roy. Sci. Liège* 6(6), 141-151 (2.4).

Cavanie, A. G. (1973). Etude non Linéaire des Ondes Internes dans un Milieu à Deux Couches Fluides Sans Rotation. Centre National pour l'Exploitation des Océans, Brest, France, Tech. Rept. 16, 11 pp. (3.10).

Cherkesov, L. V. (1973). Surface and internal waves. Translation of monograph, Poverkhnostnye i Vnutrennie Volny, Kiev, 1973, Chapters 6-7. Joint Publications Research Service, Arlington, Va., JPR6-60867, 80 pp.

Cohen, J. K. and N. Bleistein (1973). "Ray Method Formalism for Internal Waves in a Deep Ocean: Propogation, Initiation, Reflection." Div. of Math. Sciences, Denver Research Inst., Colo., Rept. MS-R-7405, 38 pp.

Crew, H. and N. Plutchak (1974). Time varying rotary spectra. *J. Oceanogr. Soc. Japan* 30(2), 61-66 (1.10, 3.7).

Csanady, G. T. (1973). Transverse internal seiches in large oblong lakes and marginal seas. *J. Phys. Oceanogr.* 3(4), 439-447 (3.8).

Csanady, G. T. and J. T. Scott (1974). Baroclinic coastal jets in Lake Ontario during IFYGL. *J. Phys. Oceanogr.* 4(4), 524-541 (1.5).

Curtin, T. B. and C. N. K. Mooers (1975). Observation and interpretation of a high-frequency internal wave packet and surface slick pattern. *J. Geophys. Res.* 80(6), 882-894 (1.3, 1.4).

Darbyshire, J. and A. Edwards (1973). Internal waves in Llyn Tegid. *Pure Appl. Geophys.* 111(10), 2359-2394 (1.5, 2.3).

Datta, R. N. (1974). Atmospheric gravity waves induced by earth's diurnal rotation. *J. Geophys. Res.* 79(10), 1583-1584 (2.1).

DeFerrari, H. A. (1974). Effects of horizontally varying internal wavefields on multipath interference for propagation through the deep sound channel. *J. Acoust. Soc. Am.* 56(1), 40-46 (1.9).

Delisi, D. P. and I. Orlanski (1975). On the role of density jumps in the reflection and breaking of internal gravity waves. *J. Fluid Mech.* (in press).

Desaubies, Y. J. F. (1973). Internal waves near the turning point. *Geophys. Fluid Dyn.* 5, 143-154 (3.3, 3.4).

Desaubies, Y. J. F. (1975). A linear theory of internal wave spectra and coherences near the Väisälä frequency. *J. Geophys. Res.* 80(6), 895-899 (3.3).

Dore, B. D. (1973). On mass transport induced by interfacial oscillations at a single frequency. *Proc. Camb. Phil. Soc.* 74, 333-347 (4.11).

Drazin, P. G. (1974). Kelvin-Helmholtz instability of a slowly varying flow. *J. Fluid Mech.* 65(4), 781-797 (4.3).

Fofonoff, N. P. (1973). Description of ocean currents (Review of recent work on deep ocean currents). *Oceanology* 13(1), 152-157 (1.4).

Fomin, L. M. (1973). Inertial oscillations in a horizontally inhomogeneous current velocity field. *Izv. Atmos. Ocean. Phys.* 9, 37-40 (3.7).

Forrester, W. D. (1974). Internal tides in St. Lawrence Estuary. *J. Mar. Res.* 32(1), 55-66 (1.5).

Francis, S. H. (1973). Acoustic-gravity modes and large-scale traveling ionospheric disturbances of a realistic, dissipative atmosphere. *J. Geophys. Res.* 78(13), 2278-2301 (3.4).

Francis, S. H. (1973). Lower-atmospheric gravity modes and their relation to medium-scale traveling ionospheric disturbances. *J. Geophys. Res.* 78(34), 8289-8295 (3.4).

Francis, S. H. (1974). A theory of medium-scale traveling ionospheric disturbances. *J. Geophys. Res.* 79(34), 5245-5260 (3.4).

Francis, S. H. (1975). Global propagation of atmospheric gravity waves: a review. *J. Atmos. Terr. Phys.* (in press) (3.4).

Frankignoul, C. J. (1972). Stability of finite amplitude internal waves in a shear flow. *Geophys. Fluid Dyn.* 4, 91-99 (4.7).

Frankignoul, C. J. (1974). Observed anisotropy of spectral characteristics of internal waves induced by low-frequency currents. *J. Phys. Oceanogr.* 4(4), 625-634 (4.7).

Frankignoul, C. J. (1974). Preliminary observations of internal wave energy flux in frequency, depth-space. *Deep-Sea Res.* 21, 895-909 (1.5, 1.10).

Frankignoul, C. J. (1974). A cautionary note on the spectral analysis of short internal wave records. *J. Geophys. Res.* 74(24), 3459-3462 (1.10).

Garrett, C. J. R. (1973). The effect of internal wave strain on vertical spectra of fine-structure. *J. Phys. Oceanog.* 3(1), 83-85 (4.12).

Garrett, C. J. R. and W. H. Munk (1973). Internal wave breaking and microstructure (The chicken and the egg). *Boundary-Layer Meteorology* 4, 37-45 (4.1, 4.12).

Garrett, C. J. R. and W. H. Munk (1975). Space-time scales of internal waves: a progress report. *J. Geophys. Res.* 80(3), 291-297 (1.1, 3.1, 3.4).

Gascard, J.-C. (1973). Vertical motions in a region of deep water formation. *Deep-Sea Res.* 20, 1011-1027 (1.5).

Gossard, E. E. (1974). Dynamic stability of an isentropic shear layer in a statically stable medium. *J. Atmos. Sci.* 31(2), 483-492 (3.9).

Gossard, E. E. and W. H. Hooke (1974). *Waves in the atmosphere.* Amsterdam: Elsevier Scientific Publ. Co., approx. 480 pp.

Gossard, E. E. and W. B. Sweezy (1974). Dispersion and spectra of gravity waves in the atmosphere. *J. Atmos. Sci.* 31(6), 1540-1548 (3.4).

Gould, W. J., W. J. Schmitz, Jr., and C. Wunsch (1974). Preliminary field results for a mid-ocean dynamics experiment (MODE-0), *Deep-Sea Res.* 21(11), 911-932 (1.5).

Gould, W. J. and W. D. McKee (1973). Vertical structure of semi-diurnal tidal currents in the Bay of Biscay. *Nature* 244(5411), 88-91 (1.5).

Graham, E. W. (1973). Transient internal waves produced by a moving body in a tank of density-stratified fluid. *J. Fluid Mech.* 61(3), 465-480 (1.7).

Grimshaw, R. H. J. (1974). "Internal Gravity Waves: Critical Layer Absorption in a Rotating Fluid." School of Math. Sciences, Univ. Melbourne, Australia, Res. Rept. 35, 36 pp. (4.7).

Grimshaw, R. H. J. (1974). "Nonlinear Internal Gravity Waves in a Rotating Fluid." Math. Dept., Univ. Melbourne, Australia, Res. Rept. 40 (3.10).

Grimshaw, R. H. J. (1974). "A Note on the β-Plane Approximation." School of Math. Sciences, Univ. Melbourne, Australia, Res. Rept. 20, 18 pp. (4.6).

Hachmeister, L. E. (1973). "Observations of Resonant Internal Wave Interactions over a Range of Driving Frequencies." Dept. Oceanogr., Univ. Washington, Seattle, Tech. Rept. 228, 66 pp. (2.2, 4.2).

Hachmeister, L. E. and S. Martin (1974). An experimental study of the resonant instability of an internal wave of mode 3 over a range of driving frequencies. *J. Phys. Oceanogr.* 4(3), 337-348 (2.2, 4.2).

Hachmeister, L. E. and S. J. Prinsenberg (1974). "Laboratory Investigation of Internal Wave Generation Models." Dept. Oceanogr., Univ. Washington, Seattle, Tech. Rept. 320, 57 pp. (2.2, 4.2).

Halpern, D. (1974). Observations of the deepening of the wind-mixed layer in the Northeast Pacific Ocean. *J. Phys. Oceanogr.* 4, 454-466 (1.5, 2.3).

Hayes, S. P. (1975). Preliminary measurements of the time-lagged coherence of vertical temperature profiles. *J. Geophys. Res.* 80(3), 307-311 (1.4, 1.5).

Hayes, S. P., T. M. Joyce, and R. C. Millard, Jr. (1975). Measurements of vertical fine structure in the Sargasso Sea. *J. Geophys. Res.* 80(3), 314-319 (4.12).

Hector, D., J. Cohen, and N. Bleistein (1972). Ray method expansions for surface and internal waves in inhomogeneous oceans of variable depth. *Stud. Appl. Math.* 51(2), 121-137 (4.4).

Hendershott, M. C. (1973). Inertial oscillations of tidal period. In, *Progress in oceanography*, Vol. 6, B. A. Warren (Ed.), Oxford: Pergamon Press, pp. 1-27 (3.7).

Hendry, R. (1974). Phase stable baroclinic tides in the Sargasso Sea. *EOS Trans. Amer. Geophys. Un.* 56(12), 1138 (abstract).

Hines, C. O. (1973). A critique of multilayer analyses in application to the propagation of acoustic-gravity waves. *J. Geophys. Res.* 78(1), 265-273 (3.4).

Hines, C. O. (1973). Reply. *J. Geophys. Res.* 78(10), 1735-1736.

Hines, C. O. (1974). Propagation velocities and speeds in ionospheric waves: a review. *J. Atmos. Terr. Phys.* 36, 1179-1204 (3.4).

Hines, C. O. and colleagues (1974). *The upper atmosphere in motion: a selection of papers with annotation.* AGU Monograph 18. Washington, D.C.: Amer. Geophys. Un. Press, approx. 1000 pp.

Hollan, E. (1974). Wenn der Bodensee aufgewühlt wird. *Umschau* 74(5), 152-154 (1.5, 2.3, 3.8).

Hughes, B. A. (1973). Nonlinear resonant inertial wave interactions. *Phys. Fluids* 16, 1805-1809 (3.7, 4.2).

Hughes, B. A. and H. L. Grant (1974). "The Interaction of Wind Waves and Internal Waves: Experimental Measurements." Defence Research Establishment Pacific, Victoria, B. C., Canada, Rept. 74-1, 109 pp. (1.4, 4.2, 4.10).

Huppert, H. E. and M. E. Stern (1974). Ageostrophic effects in rotating stratified flow. *J. Fluid Mech.* 62(2), 369-385 (2.1).

Ichiye, T. (1973). Ocean microstructure due to instability for different heat-salt diffusivity. *Mém. Soc. Roy. Sci. Liège* 6(4), 69-80 (4.12).

Ivanov, Yu. A. and Ye. G. Morozov (1974). Deformation of internal gravity waves by a stream with horizontal velocity shift (in Russian, with English summary). *Okeanologia* 14(3), 457-461 (4.7).

Jones, W. L. (1973). Asymmetric wave stress tensors and wave spin. *J. Fluid Mech.* 58, 737-748 (3.6).

Jones, W. L. and D. D. Houghton (1972). The self-destructing internal gravity wave. *J. Atmos. Sci.* 29, 844-849 (4.7).

Joyce, T. M. (1974). Fine-structure contamination of moored temperature sensors: a numerical experiment. *J. Phys. Oceanogr.* 4(2), 183-190 (4.12).

Joyce, T. M. (1974). Nonlinear interactions among standing surface and internal gravity waves. *J. Fluid Mech.* 63(4), 801-825 (1.8, 4.2).

Joyce, T. M. (1974). Vertical velocity of internal waves from a tri-mooring. *EOS Trans. Amer. Geophys. Un.* 55(4), 296 (abstract) (1.2).

Kamenkovich, V. M. and A. B. Odulo (1972). Free oscillations in a stratified compressible ocean of constant depth. *Izv. Atmos. Ocean. Phys.* 8, 693-700 (3.4).

Kamykowski, D. (1974). Possible interactions between phytoplankton and semidiurnal internal tides. *J. Mar. Res.* 32(1), 67-89 (1.9).

Kanari, S. (1974). Large scale motions in Lake Biwa. *Proc. Japan Academy* 50(8), 628-633 (1.5, 3.8).

Kanari, S. (1974). On the study of numerical experiments of two-layered Lake Biwa. *Japan. Jour. Limnol.* 35(1), 1-17 (3.8).

Kanari, S. (1974). Some results of observations of the long-period internal seiches in Lake Biwa. *Japan. Jour. Limnol.* 35(4), 137-148 (1.5, 3.8).

Kanari, S., N. Imasato, and H. Kunishi (1974). On the studies of internal waves in Lake Biwa (IV): an approximate analytical solution of the rotating internal seiches (in Japanese). *Annals, Disaster Prevention Res. Inst., Kyoto Univ., Japan* 17B, 247-256 (3.8).

Karabasheva, E. I., N. G. Kozhelupova, Yu. Z. Miropol'skii, V. T. Paka, and B. N. Filyushkin (1974). Some data on the spatial structure of the internal wave field in the ocean (in Russian, with English summary). *Okeanologia* 14(3), 462-467 (1.4, 1.5).

Katz, E. J. (1973). Profile of an isopycnal surface in the main thermocline of the Sargasso Sea. *J. Phys. Oceanogr.* 3(4), 448-457 (1.5).

Katz, E. J. (1975). Tow spectra from MODE. *J. Geophys. Res.* 80 (in press) (1.4, 1.5).

Keller, J. B. and M. C. Shen (1975). Uniform asymptotic expansion for wave propagation in a rotating compressible fluid of variable depth. *SIAM J. Appl. Math.* 28 (in press) (3.10, 4.4).

Kelley, J. C. (1973). Effects of internal waves on vertical distributions of non-conservative properties in an upwelling area. *EOS Trans. Amer. Geophys. Un.* 54(11), 1117 (abstract) (1.9).

Keunecke, K.-H. (1973). On the observation of internal tides at the continental slope off the coast of Norway. *„Meteor" Forsch.-Ergebnisse A, Meteorol. Hydrol.* 12, 24-36 (1.5).

Keunecke, K.-H. and L. Magaard (1975). Measurements by means of towed thermistor cables and problems of their interpretation with respect to mesoscale processes. *Mém. Soc. Roy. Sci. Liège* (in press) (1.5, 1.10).

Kielmann, J., W. Krauss, and K.-H. Keunecke (1973). Currents and stratification in the Belt Sea and the Arkona Basin during 1962-1968. *Kiel. Meeresforsch.* 29(2), 90-111 (1.4, 1.5, 3.8).

Kindle, J. C. (1974). The horizontal coherence of inertial oscillations in a coastal region. *Geophys. Res. Letters* 1(3), 127-130 (1.5).

Kitaygorodskii, S. A., Yu. Z. Miropol'skii, and B. N. Filyushkin (1973). Use of ocean temperature fluctuation data to distinguish internal waves from turbulence. *Izv. Atmos. Ocean. Phys.* 9(3), 149-159 (1.4, 1.5).

Klemp, J. B. and D. K. Lilly (1975). The dynamics of wave-induced downslope winds. *J. Atmos. Sci.* (in press) (3.9).

Klostermeyer, J. (1973). Comments on a paper by C. I. Hines, a Critique of multilayer analyses in application to the propagation of acoustic-gravity waves. *J. Geophys. Res.* 78(10), 1733-1734 (3.4).

Konyayev, K. V. and K. D. Sabinin (1973). New data on internal waves in the ocean obtained with distributed temperature sensors. *Doklady, Earth Science Sections* 209, 3-8; translated from *Doklady Akad. Nauk SSSR* 209(1), 86-89 (1.4).

Konyayev, K. V. and K. D. Sabinin (1973). A resonator theory of generation of internal waves in the ocean. *Doklady, Earth Science Sections* 210, 36-39; translated from *Doklady Akad. Nauk SSSR* 210(6), 1342-1345 (2.1).

Krauss, W., P. Koske, and J. Kielmann (1973). Observations on scattering layers and thermoclines in the Baltic Sea. *Kiel. Meeresforsch.* 24, 85-89 (1.4, 1.8, 1.9).

Lacombe, H. (1972). Les principaux problèmes relatifs aux ondes internes d'après les exposés annoncés. *Rapports et Procès-verbaux, Intern. Council Expl. Sea (Copenhagen)* 162, 45-52.

Lazier, J. R. N. (1973). Temporal changes in some fresh water temperature structures. *J. Phys. Oceanogr.* 3(2), 226-229 (4.12).

Leaman, K. D. and T. B. Sanford (1975). Vertical energy propagation of inertial waves: a vector spectral analysis of velocity profiles. *J. Geophys. Res.* 80 (in press).

Lee, C.-Y. and R. C. Beardsley (1974). The generation of long nonlinear internal waves in a weakly stratified shear flow. *J. Geophys. Res.* 79(3), 453-462 (1.3, 1.7, 2.5, 3.10).

Leonov, A. I. and Yu. Z. Miropol'skii (1973). Resonant excitation of internal gravity waves in the ocean by atmospheric-pressure fluctuations. *Izv. Atmos. Ocean. Phys.* 9(8), 480-485 (2.3).

Levkov, M. P. and L. M. Cherkesov (1973). Boundary surface layers and internal long waves. *Izv. Atmos. Ocean. Phys.* 9(3), 165-168 (4.9).

Lewis, J. E., B. M. Lake, and D. R. S. Ko (1974). On the interaction of internal waves and surface gravity waves. *J. Fluid Mech.* 63(4), 773-800, 1 plate (2.2).

Lilly, D. K. and P. J. Kennedy (1973). Observations of a stationary mountain wave and its associated momentum flux and energy dissipation. *J. Atmos. Sci.* 30(6), 1135-1152 (3.6, 3.9).

Lilly, D. K. and P. F. Lester (1974). Waves and turbulence in the stratosphere. *J. Atmos. Sci.* 31(3), 800-812 (3.9).

Linden, P. F. and J. S. Turner (1975). Small scale mixing in stably stratified fluids: a report on Euromech 51. *J. Fluid Mech.* 67(1) 1-16 (1.9, 4.11).

Lindzen, R. S., L. N. Howard, A. D. McEwan, D. P. DeLisi, and E. E. Gossard (1973). Hydrodynamics of stratified fluids (the applicability of linear theory). *Boundary-Layer Meteorology* 5, 227-231 (2.2, 3.10).

Long, R. R. (1972). Turbulence from breaking of internal gravity waves. *Rapports et Procès-Verbaux, Intern. Council Expl. Sea (Copenhagen)* 162, 13-18.

Long, R. R. (1974). Some experimental observations of upstream disturbances in a two-fluid system. *Tellus* 26(3), 313-317 (1.7, 3.9).

Lovorn, F. T. (1973). "An Investigation of Mean Temperature Structure and Internal Waves in Puget Sound Central Basin." Dept. of Oceanogr, Univ. Washington, Seattle, TR-287, 74 pp.

Lozovatskii, I. D. and S. M. Shapovalov (1973). Determination of certain characteristics of internal waves at a given Väisälä frequency profile. *Izv. Atmos. Ocean. Phys.* 9(4), 248-250 (3.4).

McEwan, A. D. and R. M. Robinson. Parametric instability of internal gravity waves. *J. Fluid Mech.* (in press) (1.8, 4.2).

McKean, R. S. (1974). Interpretation of internal wave measurements in the presence of fine-structure. *J. Phys. Oceanogr.* 4(2), 200-213 (4.12).

McKean, R. S. and T. E. Ewart (1974). Temperature spectra in the deep ocean off Hawaii. *J. Phys. Oceanogr.* 4(2), 191-199 (1.4, 1.5).

Magaard, L. (1974). On the generation of internal gravity waves by meteorological forces. *Mém. Soc. Roy. Sci. Liège* 6(6), 79-85 (2.3).

Makshtas, Ya. P. and K. D. Sabinin (1972). Relation between variations in the depth of sound-scattering layers and internal waves in the ocean. *Oceanology* 12(4), 625-628 (1.9).

Maslowe, S. A. (1972). The generation of clear air turbulence by nonlinear waves. *Stud. Appl. Math.* 51(1), 1-16.

Metcalf, J. I. (1974). Ph.D. thesis. "On the Origin and Structure of Radar Echo Layers in the Clear Atmosphere." Dept. Geophys. Sci., Univ. Chicago, 244 pp.

Metcalf, J. I. (1975). Microstructure of radar echo layers in the clear atmosphere. *J. Atmos. Sci.* 32(2), 362-370 (1.8).

Metcalf, J. I, (1975). Gravity waves in a low-level inversion. *J. Atmos. Sci.* 32(2), 351-361 (1.8).

Metcalf, J. I. and D. Atlas (1973). Microscale ordered motions and atmospheric structure associated with thin echo layers in stably stratified zones. *Boundary-Layer Meteorology* 4, 7-35 (1.8).

Meyer, R. E. (1973). Note on lee waves and windward modes. *J. Geophys. Res.* 78(33), 7917-7918 (3.9).

Mied, R. P. (1972). Ph.D. thesis. "Stability of Plane Internal Waves in a Uniformly Stratified Fluid." The Johns Hopkins Univ., Baltimore, Md.

Mied, R. P. and J. P. Dugan (1974). Internal gravity wave reflection by a layered density anomaly. *J. Phys. Oceanogr.* 4(3), 493-498 (4.1).

Miles, J. W. (1974). On Laplace's tidal equations. *J. Fluid Mech.* 66(2), 241-260 (3.2).

Miropol'skii, Yu. Z. (1973). Probability distribution of certain characteristics of internal waves in the ocean. *Izv. Atmos. Ocean. Phys.* 9(4), 226-230 (1.4).

Mooers, C. N. K. (1973). A simple model for the topographic scattering of inertial-internal waves. *EOS Trans. Amer. Geophys. Un.* 54(11), 1121 (abstract) (4.4).

Mooers, C. N. K. (1974). An indirect method for calculating the mean square bottom velocity produced by inertial-internal waves propagating over sloping boundaries. *EOS Trans. Amer. Geophys. Un.* 56(12), 1137 (abstract) (4.4).

Mooers, C. N. K. and J. F. Price (1974). Tidal and inertial motions on the West Florida Shelf: winter 1973. *EOS Trans. Amer. Geophys. Un.* 55(4), 295 (abstract) (1.5).

Mortimer, C. H. (1974). Lake hydrodynamics. *Mitt. Int. Ver. Limnol., Comm. Int. Assoc. Limnol.* 20, 124-197.

Morozov, Ye. G. (1974). Experimental studies of breaking internal waves (in Russian, with English summary). *Okeanologiya* 14(1), 25-29 (1.8).

Morozov, Ye. G. (1974). Experimental studies of temperature fluctuations in the upper layer of the West Pacific (in Russian, with English summary). *Okeanologiya* 14(4), 602-606 (1.4, 1.5).

Müller, P. (1974). "On the Interaction between Short Internal Waves and Larger Scale Motions in the Ocean. Geophysikalischen Instituten der Universität Hamburg, Hamb. Geophys. Einzelschriften Heft 23, 98 pp. (4.7).

Müller, P. and D. J. Olbers (1975). On the dynamics of internal waves in the deep ocean. *J. Geophys. Res.* 80 (in press).

Müller, P. and G. Siedler. On a diagnostic concept to test the internal wave structure (in preparation).

Munk, W. H., J. D. Woods, W. P. Birkemeier, C. G. Little, R. H. Stewart, and J. Richter (1973). Remote sensing of the ocean, *Boundary-Layer Meteorology* 5, 201-209 (1.2).

Navrotskii, V. V. (1974). Analysis of the thermal structure of the surface layer on transections in the Sea of Japan. *Oceanology* 12(1), 31-39 (1.1, 1.4).

Navrotskii, V. V., V. T. Paka, and E. I. Karabasheva (1972). Characteristics of thermal inhomogeneities on sections in the Atlantic Ocean. *Izv. Atmos. Ocean. Phys.* 8, 174-181 (1.4, 1.5).

Neshyba, S. J., V. T. Neal, and R. Tucker (1975). Rapid high-resolution *in situ* sampling of the internal wave on a density step. *J. Geophys. Res.* 80(3), 312-313 (1.2, 1.4).

Neshyba, S. J. and E. J. C. Sobey (1975). Vertical cross-coherence and cross-bispectra between internal waves measured in a multiple-layered ocean. *J. Geophys. Res.* 80 (in press).

Nesterov, S. V. (1972). Resonant interaction of surface and internal waves. *Izv. Atmos. Ocean. Phys.* 8, 252-254 (4.2).

Nihoul, J. C. J. (1972). On the energy spectra of a random field of internal waves. *Tellus* 24(2), 161-163 (4.7).

Odulo, A. B. (1973). Internal and inertial waves near an inclined shore and in a corner region. *Oceanology* 13(5), 655-658 (4.4).

Olbers, D. J. (1974). "On the Energy Balance of Small Scale Internal Waves in the Deep-Sea." Geophysikalischen Instituten der Universität Hamburg, Hamb. Geophys. Einzelschriften Heft 24, 91 pp. (3.6, 3.10, 4.2, 4.7).

Orlanski, I. (1973). Trapeze instability as a source of internal gravity waves, Part I. *J. Atmos. Sci.* 30(6), 1007-1016 (2.1).

Orlanski, I. and B. B. Ross (1973). Numerical simulation of the generation and breaking of internal gravity waves. *J. Geophys. Res.* 78(36), 8808-8826 (1.8, 4.5).

Panicker, N. N. (1974). Stability of a tripod mooring for deep-sea internal wave experiment. *EOS Trans. Amer Geophys. Un.* 55(4), 286 (abstract) (1.2).

Pao, Y.-H. (1973). Measurements of internal waves and turbulence in two-dimensional stratified shear flows. *Boundary-Layer Meteorology* 5, 177-193 (1.7, 3.9).

Perkins, F. W. and S. H. Francis (1974). Artificial production of traveling ionospheric disturbances and large-scale atmospheric motion. *J. Geophys. Res.* 79(25), 3879-3881 (2.1).

Petrie, B. (1974). Ph.D. thesis. "Surface and Internal Tides on the Scotian Shelf and Slope." Dalhousie Univ., Halifax, Nova Scotia.

Phillips, O. M. (1973). On the interactions between internal and surface waves. *Izv. Atmos. Ocean. Phys.* 9(10), 565-568 (4.2).

Phillips, O. M. (1974). Nonlinear dispersive waves. In, *Annual review of fluid mechanics*, Vol. 6. M. Van Dyke, W. G. Vincenti, and J. V. Wehausen (Eds.), Palo Alto, Calif.: Annual Reviews, pp. 93-110 (2.2, 4.2).

Pinkel, R. (1975). Upper ocean internal wave measurements from FLIP. *J. Geophys. Res.* 80 (in press).

Polyanskaya, V. A. (1974). Influence of high-frequency internal waves on the sound field on a point source in the ocean. *Sov. Phys. Acoust.* 20, 55-59 (1.9).

Porter, R. P., R. C. Spindel, and R. J. Jaffee (1974). Acoustic-internal wave interaction at long ranges in the ocean. *J. Acous. Soc. Amer.* 56(5), 1426-1436 (1.9).

Preisendorfer, R. W., J. C. Larsen, and M. Sklarz (1973). Electromagnetic fields induced by plane-parallel internal and surface ocean waves. *EOS Trans. Amer. Geophys. Un.* 54(11), 1120 (abstract) (1.9).

Prinsenberg, S. J., W. L. Wilmot, and M. Rattray, Jr. (1974). Generation and dissipation of coastal internal tides. *Deep-Sea Res.* 21(4), 263-281 (2.4, 4.4).

Proni, J. R. and J. R. Apel (1975). On the use of high-frequency acoustics for the study of internal waves and micro-structure. *J. Geophys. Res.* 80 (in press).

Proni, J. R., J. R. Apel, and H. M. Byrne (1974). An analysis of correlations existing among multisensor observations of internal waves during NYBERSEX. *EOS Trans. Amer. Geophys. Un.* 56(12), 1137 (abstract (1.2).

Rao, G. V. P. (1973). Waves generated in rotating fluids by travelling forcing effects. *J. Fluid Mech.* 61(1), 129-158 (1.7, 2.5).

Rhines, P. (1973). Observations of the energy-containing oceanic eddies, and theoretical models of waves and turbulence. *Boundary-Layer Meteorology* 4, 345-360 (1.4, 1.5).

Rigby, F. (1974). "Theoretical Calculations of Internal Wave Drag on Sea Ice." Dept. Atmos. Sci., Univ. Washington, Seattle, Tech. Rept. 26, 12 pp. (1.9).

Robinson, I. S. (1973). "Internal Tides in the British Shelf Seas." Inst. Coastal Oceanogr. and Tides, Birkenhead, Cheshire, England, Internal Rept. 28, 28 pp. (1.5).

Robinson, R. M., and A. D. McEwan. Instability of a periodic boundary layer in a stratified fluid. *J. Fluid Mech.* (in press) (1.8).

Roskes, G. J. (1972). Some inertial wave sideband instabilities. *Phys. Fluids* 15(5), 737-740 (3.7, 4.5).

Rossby, T. (1974). Studies of the vertical structure of horizontal currents near Bermuda. *J. Geophys. Res.* 79(12), 1781-1791 (1.5).

Sabinin, K. D. (1973). Certain features of short-period internal waves in the ocean. *Izv. Atmos. Ocean. Phys.* 9, 32-36 (1.4).

Saint-Guily, B. (1972). Réponse d'une mer faiblement stratifiée à une impulsion superficielle, pp. 243-248. In, Coll. Intern. du Centre National de la Recherche Scientifique, No. 215, Paris, Oct. 1972 (2.3, 4.11).

Sandstrom, H. (1973). The effect of boundary curvature on reflection of internal waves. *Mém. Soc. Roy. Sci. Liège* 4(6), 183-190.

Sanford, T. B. (1974). Observations of strong current shears in the deep ocean and some implications on sound rays. *J. Acoust. Soc. Am.* 56(4), 1118-1121 (1.9).

Sanford, T. B. (1975). Observations of the vertical structure of internal waves. *J. Geophys. Res.* 80 (in press).

Schott, F. (1973). A method to determine characteristic parameters of internal waves. *Mém. Soc. Roy. Sci. Liège* 4(6), 163-169.

Schott, F. (1974). Über die raum-zeitlich Struktur von Stromschwankungen im Meer unter besonderer Berücksichtigung barokliner Gezeiten. Inst. Meereskunde, Kiel, Germany, 147 pp.

Shen, M. C. and J. B. Keller (1973). Ray method for nonlinear wave propagation in a rotating fluid of variable depth. *Phys. Fluids* 16(10), 1565-1572 (3.10, 4.4).

Shepard, F. P., N. F. Marshall, and P. A. McLoughlin (1974). "Internal waves" advancing along submarine canyons. *Science* 183, 195-197 (1.1, 1.4, 1.5).

Shepard, F. P., N. F. Marshall, and P. A. McLoughlin (1974). Currents in submarine canyons. *Deep-Sea Res.* 21(9), 691-706 (1.5).

Sheppard, D. M. and O. H. Shemdin (1974). A multipurpose facility for studying internal waves. *EOS Trans. Amer. Geophys. Un.* 55(4), 295 (abstract) (1.7).

Shevtsov, V. P. and A. P. Volkov (1973). A method for measuring the vertical profiles of ocean currents from a ship. *Oceanology* 13(6), 916-920 (1.2).

Shonting, D. H. (1974). Current observations from the Western Mediterranean during COBLAMED 69. *Limnol. Oceanogr.* 19(5), 866-874 (1.5).

Siedler, G. (1974). The fine-structure contamination of vertical velocity spectra in the deep ocean. *Deep-Sea Res.* 21, 37-46 (4.12).

Siedler, G. (1974). Observations of internal wave coherence in the deep ocean. *Deep-Sea Res.* 21, 597-610 (1.5).

Smirnov, V. N. (1972). Ice cover oscillations resulting from internal waves of the ice ocean. *Doklady, Earth Science Sections* 206, 16-18; translated from *Doklady Akad. Nauk. SSSR* 206(5), 1106-1108 (1.5, 1.9).

Smith, R. (1973). Evolution of inertial frequency oscillations. *J. Fluid Mech.* 60(2), 383-389 (3.7).

Snodgrass, F., W. Brown, and W. H. Munk (1975). MODE: IGPP measurements of bottom pressure and temperature. *J. Phys. Oceanogr.* 5(1), 63-74 (1.2).

Spindel, R. C., R. P. Porter, and R. J. Jaffee (1974). Long-range sound fluctuations with drifting hydrophones. *J. Acoust. Soc. Am.* 56(2), 440-446 (1.9).

Stegen, G. R., K. Bryan, J. L. Held, and F. Ostapoff (1975). Dropped horizontal coherance based on temperature profiles in the upper thermocline. *J. Geophys. Res.* 80 (in press).

Stevenson, T. N. (1973). The phase configuration for internal waves around a body moving in a density stratified fluid. *J. Fluid Mech.* 60(4), 759-767, 6 plates (1.7).

Stevenson, T. N., J. N. Bearon, and N. H. Thomas (1974). An internal wave in a viscous heat-conducting isothermal atmosphere. *J. Fluid Mech.* 65(2), 315-323 (4.9).

Thomas, N. H. and T. N. Stevenson (1973). An internal wave in a viscous ocean stratified by both salt and heat. *J. Fluid Mech.* 61(2), 301-304 (4.9).

Thompson, O. E. (1973). Resonant oscillations of intermediate frequency in a stratified atmosphere. *J. Geophys. Res.* 78(27), 6173-6181 (2.1).

Thornton, E. B. and N. E. J. Boston (1974). A study of the effect of internal wave induced turbulence on small scale temperature structure in shallow water. *EOS Trans. Amer. Geophys. Un.* 55(4), 296 (abstract) (4.12).

Thorpe, S. A. (1973). Experiments on instability and turbulence in a stratified shear flow. *J. Fluid Mech.* 61(4), 731-751, 11 plates (1.8, 4.3).

Thorpe, S. A. (1974). Near-resonant forcing in a shallow two-layer fluid: a model for the internal surge in Loch Ness? *J. Fluid Mech.* 63(3), 509-527 (1.3, 1.7, 2.3, 3.5, 3.8).

Thorpe, S. A. (1975). The excitation, dissipation, and interaction of internal waves in the deep ocean. *J. Geophys. Res.* 80(3) 328-338 (2.1, 3.6, 4.1).

Titov, V. B. (1973). Rotary currents, according to measurements in the ocean. *Oceanology* 12(2), 177-181 (1.4).

Titov, V. B. (1973). Some distinctive features of meso-scale motions in the sea. *Oceanology* 13(6), 794-798 (3.7).

Tsuji, Y. and Y. Nagata (1973). Stokes' expansion of internal deep water waves to the fifth order. *J. Oceanogr. Soc. Japan* 29(2), 61-69 (3.5, 3.10).

Turner, J. S. (1973). *Buoyancy effects in fluids.* London: Cambridge Univ. Press, 367 pp. (1.6, 1.7, 1.8, 3.2, 3.4, 3.6, 3.8, 3.9, 3.10, 3.11, 4.2, 4.3, 4.4, 4.9, 4.11, 4.12).

Van Leer, J., W. Düing, R. Erath, E. Kennelly, and A. Speidel (1974). The Cyclesonde: an unattended vertical profiler for scalar and vector quantities in the upper ocean. *Deep-Sea Res.* 21(5), 385-400 (2.1).

Volland, H. (1973). Comments on a paper by C. O. Hines. *J. Geophys. Res.* 78(10), 1737-1738.

Volosov, V. M. (1974). Asymptotic analysis of a certain type of gravitational-gyroscopic internal wave (in Russian, with English summary). *Okeanologiya* 14(4), 589-594 (3.4).

Webster, F. (1972). Estimates of the coherence of ocean currents over vertical distances. *Deep-Sea Res.* 19, 35-44 (1.5).

Weinberg, N. L., J. G. Clark, and R. P. Flanagan (1974). Internal tidal influence on deep-ocean acoustic-ray propagation. *J. Acoust. Soc. Am.* 56(2), 447-458 (1.9).

Williams, K. G. and T. H. Bell, Jr. (1972). Cyclic flows resulting from localized disturbances in stratified fluids, pp. unknown (typed draft is 9 pages long). In, Intern. Symp. Stratified Flows, Novosibirsk, 1972 (2.5, 4.11).

Winant, C. D. (1974). Internal surges in coastal waters. *J. Geophys. Res.* 79(30), 4523-4526 (1.3, 1.4, 1.8).

Woods, J. D. (1973). Space-time characteristics of turbulence in the seasonal thermocline. *Mém. Soc. Roy. Sci. Liège* 6(6), 109-130 (4.11).

Wunsch, C. I. (1975). Deep ocean internal waves: what do we really know? *J. Geophys. Res.* 80(3), 339-343 (1.2, 3.4, 3.6).

Wunsch, C. I. (1975). Internal tides in the ocean. *Rev. Geophys.* (in press).

Yampol'skii, A. D. (1973). Certain energy-balance features of intertial oscillations in the velocity field in the ocean. *Izv. Atmos. Ocean. Phys.* 9(11), 695-697 (3.7).

Yeh, K. C. and C. H. Liu (1974). Acoustic-gravity waves in the upper atmosphere. *Rev. Geophys. Space Phys.* 12(2), 193-216 (2.1, 3.2, 3.9, 4.6).

Yih, C.-S. (1974). Instability of stratified flows as a result of resonance. *Phys. Fluids* 17(8), 1483-1488 (3.9, 4.2).

Yih, C.-S. (1974). Progressive waves of permanent form in continuously stratified fluids. *Phys. Fluids* 17(8), 1489-1495 (3.4, 3.11).

Yih, C.-S. (1974). Wave motion in stratified fluids. In, *Nonlinear waves*, S. Leibovich and A. R. Seebass (Eds.). Ithaca, N.Y.: Cornell Univ. Press, pp. 263-290 (3.9).

Zenk, W. and M. G. Briscoe (1974). "The Cape Cod Experiment on Near-Surface Internal Waves." Woods Hole Oceanogr. Inst., WHOI Tech. Rept. 74-87, 52 pp. (1.4).

Zenk, W. and E. J. Katz (1975). On the stationarity of temperature spectra at high horizontal wave numbers. *J. Geophys. Res.* 80 (in press).

BIBLIOGRAPHY

Afashagov, M. S. (1969). Internal waves in a non-homogeneous atmosphere. *Izv. Atmos. Ocean. Phys.* 5(5), 249-251.

Allan, T. D. (1966). "Observed Sound Velocity and Temperature Profiles in the Strait of Gibraltar during the Passage of an Internal Wave." NATO SACLANT ASW Research Centre, La Spezia, Italy, Tech. Rept. 66, DDC AD 802 091, 30 pp.

Arthur, R. S. (1954). Oscillations in sea temperature at Scripps and Oceanside piers. *Deep-Sea Res.* 2, 107-121.

Arthur, R. S. (1960). Variation of sea temperature off La Jolla. *J. Geophys. Res.* 65(12), 4081-4086.

Atlas, D. and J. I. Metcalf (1970). "The Amplitude and Energy of Breaking Kelvin-Helmholtz Waves and Turbulence." Laboratory for Atmospheric Probing, Univ. Chicago, Tech. Rept. 19, 20 pp.

Atlas, D., J. I. Metcalf, J. H. Richter, and E. E. Gossard (1970). The birth of CAT and microscale turbulence. *J. Atmos. Sci.* 27(6), 903-913.

Baines, P. G. (1971a). The reflexion of internal/inertial waves from bumpy surfaces. *J. Fluid Mech.* 46(2), 273-291.

Baines, P. G. (1971b). The reflexion of internal/inertial waves from bumpy surfaces. Part 2. Split reflexion and diffraction. *J. Fluid Mech.* 49(1), 113-131.

Baines, P. G. (1973). The generation of internal tides by flat-bumps topography. *Deep-Sea Res.* 20, 197-205.

Baines, P. G. (1974). The generation of internal tides over steep continental slopes. *Phil. Trans. Roy. Soc., Ser. A* 277(1263), 27-58.

Baker, D. J., Jr. (1972). Viscous overturning - a mechanism for oceanic microstructure? *EOS Trans. Amer. Geophys. Un.* 53(4), 415 (abstract).

Ball, F. K. (1964). Energy transfer between external and internal gravity waves. *J. Fluid Mech.* 19(3), 465-478.

Barcilon, A., S. Blumsack, and J. Lau (1972). Reflection of internal gravity waves by small density variations. *J. Phys. Oceanogr.* 2, 104-107.

Batchelor, G. K. (1970). *An introduction to fluid dynamics.* London: Cambridge Univ. Press, 615 pp.

Bell, T. H., Jr. (1971). "Numerical Calculations of Dispersion Relations for Internal Gravity Waves." Naval Research Lab., Washington, D.C., NRL Rept. 7294, 46 pp.

Bell, T. H., Jr. (1973a). Ph.D. thesis. "Internal Wave Generation by Deep Ocean Flows over Abyssal Topography." The Johns Hopkins Univ., Baltimore, Md., 130 pp.

Bell, T. H., Jr. (1973b). "Internal Wave Generation by Submerged Bodies: Mean Flow Effects." Naval Research Lab., Washington, D.C., NRL Memo Rept. 2553, 34 pp.

Bell, T. H., Jr. (1973c). On the scattering of internal waves by deep ocean fine-structure. *J. Phys. Oceanogr.* 3, 239-241.

Bell, T. H., Jr. (1974a). Effects of shear on the properties of internal gravity wave modes. *Dtsch. Hydrogr. Zeit.* 27, 57-62.

Bell, T. H., Jr. (1974b). Vertical mixing in the deep ocean. *Nature* 251, 43-44.

Belshé, J. C. (1968). Internal waves and the sound velocity structure in waters of the Kaulakahi Channel, Hawaii. *Trans. Amer. Geophys. Un.* 49(1), 211 (abstract).

Belyakov, Yu. M. and O. M. Belyakova (1963). Determination of some parameters of internal waves in the Sargasso Sea by the autocorrelation method. *Soviet Oceanogr.* 2, 1-6.

Benjamin, T. B. (1966). Internal waves of finite amplitude and permanent form. *J. Fluid Mech.* 25, 241-270.

Benjamin, T. B. (1967). Internal waves of permanent form in fluids of great depth. *J. Fluid Mech.* 29, 559-592.

Benjamin, T. B. and M. J. Lighthill (1954). On cnoidal waves and bores. *Proc. Roy. Soc. London, Ser. A* 224, 448-460.

Bernstein, R. L. and K. Hunkins (1971). Inertial currents from a three-dimensional array in the Arctic Ocean. *EOS Trans. Amer. Geophys. Un.* 52(4), 253 (abstract).

Bird, R. B., W. E. Stewart, and E. N. Lightfoot (1960). *Transport phenomena.* New York: J. Wiley and Sons, Inc., 780 pp.

Birge, E. A. (1909). On the evidence for temperature seiches. *Trans. Wisc. Acad. Sci. Arts Lett.* 16, 1005-1016.

Blumen, W. and R. C. Hendl (1969). On the role of Joule heating as a source of gravity-wave energy above 100 km. *J. Atmos. Sci.* 26(2), 210-217.

Bockel, M. (1962). Travaux océanographiques de l' "Origny" à Gibraltar. Campagne Internationale 15 Mai-15 Juin 1961. 1. Partie: Hydrologie dans le détroit. *Cah. Océanogr.* 14, 325-329.

Booker, J. R. and F. P. Bretherton (1967). Critical layer for internal gravity waves in a shear flow. *J. Fluid Mech.* 27, 513-539.

Boris, L. I. (1970a). *Atlas of the variabilities of ocean-atmosphere system of the north Atlantic*, Vol. 6 (in Russian). Leningrad: Hydrometeorological Inst., 101 pp.

Boris, L. I. (1970b). Space and time variations of tidal mixing caused by internal waves. *Oceanology* 10(5), 614-617.

Boris, L. I. (1972). Experience in the calculation of the seasonal variability of the tidal internal waves in the North Atlantic based on the mean many-year density distribution. *Rapports et Procès-Verbaux Intern. Council Expl. Sea (Copenhagen)* 162, 99-102.

Boston, N. E. J. (1963). Master's thesis. "The Internal Tide off Panama City, Florida." Texas A & M Univ., 75 pp.

Boston, N. E. J. (1964). "Observations of Tidal Periodic Internal Waves over a Three-Day Period off Panama City, Florida." Texas A & M Univ., A & M Project 286-D, Reference 64-20T, 31 August 1964, 49 pp.

Boussinesq, J. (1903). *Théorie analytique de la chaleur*, Vol. 2. Paris: Gauthier-Villars, 625 pp. (Cited by Spiegel and Veronis, 1960).

Breeding, R. J. (1972). A nonlinear model of the break-up of internal gravity waves due to their exponential growth with height. *J. Geophys. Res.* 77(15), 2681-2692.

Bretherton, F. P. (1966). Propagation of groups of internal gravity waves in a shear flow. *Quart. J. Roy. Met. Soc.* 92, 466-480.

Bretherton, F. P. (1969). Momentum transport by gravity waves. *Quart. J. Roy. Met. Soc.* 95(404), 213-243.

Bretherton, F. P. (1970). The general linearised theory of wave propagation, pp. 61-102. In, *Mathematical Problems in the Geophysical Sciences; Lectures in Applied Mathematics*, Vol. 13, W. G. Reid (Ed.), Providence, R. I.: Am. Math. Soc., pp. 61-102.

Bretherton, F. P. (1971). Introduction to internal waves, pp. 9-13. In, Lecture Notes, NATO Advanced Study Institute on Topics in Geophysical Fluid Dynamics held at the Univ. College of North Wales in Bangor, 29 June - 16 July, 1971.

Bretherton, F. P. and C. J. R. Garrett (1968). Wavetrains in inhomogeneous moving media. *Proc. Roy. Soc. London, Ser. A* 302, 529-554.

Brown, A. L., E. L. Corton, and L. S. Simpson (1955). "Power Spectrum Analysis of Internal Waves from Operation Standstill." Hydrographic Office, Washington, D.C., Tech. Rept. 26, 20 pp.

Burnside, W. (1889). On the small wave-motions of a heterogeneous fluid under gravity. *Proc. London Math. Soc.* 20, 392-397.

Byshev, V. I., Yu. A. Ivanov, and Ye. G. Morozov (1971). Study of temperature fluctuations in the frequency range of internal gravity waves. *Izv. Atmos. Ocean. Phys.* 7(1), 25-30.

Cacchione, D. A. (1970). Ph.D. thesis. "Experimental Study of Internal Gravity Waves over a Slope." MIT, and Woods Hole Oceanogr. Inst., WHOI Rept. 70-6, 226 pp.

Cairns, J. L. (1967). Asymmetry of internal tidal waves in shallow coastal waters. *J. Geophys. Res.* 72(14), 3563-3565.

Cairns, J. L. (1968). Thermocline strength fluctuations in coastal waters. *J. Geophys. Res.* 73(8), 2591-2595.

Cairns, J. L. and E. C. LaFond (1966). Periodic motions of the seasonal thermocline along the Southern California coast. *J. Geophys. Res.* 71(6), 3903-3915.

Caloi, P. and M. Migani (1964). Sulla natura fisica delle onde interne del lago di Bracciano. *Ann. Geof. Rome* 17(2), 213-220. (Reference supplied by the Deutsches Hydrographisches Inst., Hamburg, Germany.)

Caloi, P., M. Migani, and G. Pannocchia (1961). Ancora sulle onde interne del lago di Bracciano e sui fenomeni ad esse collegati. *Ann. Geof. Rome* 14(3), 347-355.

Carsola, A. J. (1967a). "Deep Water Internal Waves Study. Final Report." Lockheed, San Diego, Lockheed Rept. 20848, 44 pp.

Carsola, A. J. (1967b). "Temperature Fluctuations in the Waters Adjacent to San Clemente Island, California." Lockheed Oceanics Div., San Diego, Rept. 20474, 15 pp.

Carsola, A. J. and E. B. Callaway (1962). Two short-period internal wave frequency spectra in the sea off Southern California. *Limnol. Oceanogr.* 7(2), 155-120.

Carsola, A. J., D. P. Hamm, and J. C. Roque (1965). Spectra of temperature fluctuations over the continental borderland off Southern California. *Deep-Sea Res.* 12, 685-691.

Carsola, A. J. and C. H. Jeffress (1968). Temperature fluctuations in the Santa Catalina Basin, pp. 283-301. In, Proc. of Ocean Test Ranges and Instrumentation Conf., Honolulu, Hawaii.

Cartwright, D. E. (1959). On submarine sand-waves and tidal lee-waves. *Proc. Roy. Soc. London, Ser. A* 253, 218-241.

Cavanie, A. G. (1969). Sur la genèse et la propagation d'ondes internes dans un milieu à deux couches. *Cah. Océanogr.* 21(9), 831-843.

Cavanie, A. G. (1971). Modèle non linéaire et non hydrostatique appliqué à l'étude des fronts internes dans le détroit de Gibraltar. *Cah. Océanogr.* 23(7), 611-627.

Cavanie, A. G. (1972). Observations de fronts internes dans le détroit de Gibraltar pendant la campagne océanographique OTAN 1970 et interprétation des résultats par un modèle mathématique. *Mém. Soc. Roy. Sci. Liège* 2(6), 27-41.

Chapman, S. and J. Bartels (1940). *Geomagnetism.* Oxford: Clarendon Press, 2 vols., 1049 pp.

Chapman, S. and R. S. Lindzen (1970). *Atmospheric tides.* Dordrecht, Netherlands: D. Reidel, 200 pp.

Charnock, H. (1965). A preliminary study of the directional spectrum of short-period internal waves, pp. 177-178. In, Proc. 2nd U.S. Navy Symp. Military Oceanogr., May 5-7, 1965, Vol. 1.

Chimonas, G. (1970). Internal gravity-wave motions induced in the earth's atmosphere by a solar eclipse. *J. Geophys. Res.* 75(28), 5545-5551.

Chindonova, Yu. G. and V. A. Shulepov (1965). Sound-scattering layers as indicators of internal waves in the ocean. *Acad. Sci. USSR* 5(6), 78-81.

Chunchuzov, Ye. P. (1971). The interaction of internal waves with the mean wind in the upper atmosphere. *Izv. Atmos. Ocean. Phys.* 7(10), 719-720.

Clarke, R. A. and R. F. Reiniger (1973). A small current meter array near 39°25'N, 49°50'W. *EOS Trans. Amer. Geophys. Un.* 54(4), 311 (abstract).

Constant, F. W. (1954). *Theoretical physics.* Reading, Mass.: Addison-Wesley, 281 pp.

Cornell Aeronautical Lab. (1969). Lake motions, pp. 24-34. In, Cayuga Lake and Bell Station. Physical Effects, Final Rept. Summary. Cornell Univ., Buffalo, New York, Cal No. VT-2616-0-1.

Covez, L. (1971). Mountain waves in a turbulent atmosphere. *Tellus* 23(1), 104-110.

Cox, C. S. (1960). Moored station type C, pp. 16-19. In, Scripps Tuna Oceanogr. Res. (STOR) Program, Final Rept., Scripps Inst. Oceanogr. Ref. 60-50.

Cox, C. S. (1962). Internal waves. In, *The sea*, Vol. 1, M. N. Hill (Ed.). New York: Interscience, pp. 752-763.

Cox, C. S. (1966). Energy in semidiurnal internal waves, pp. 69-80. In, Trans. IAPO Symposium, Moscow, 25-28 May, 1966.

Cox, C. S. and H. Sandstrom (1962). Coupling of internal and surface waves in water of variable depth. *J. Oceanogr. Soc. Japan,* 20th Ann. Vol., 499-513.

Crampin, D. J. and B. D. Dore (1970). Numerical comparisons of the damping of internal gravity waves in stratified fluids. *Pure Appl. Geophys.* 79(2), 53-65.

Crapper, G. D. (1967). Ship waves in a stratified ocean. *J. Fluid Mech.* 29(4), 667-672.

Crepon, M. (1971). Ph.D. thesis. "Hydrodynamique Marine en Régime Impulsionnel." Physical Sci. Univ. Paris, 101 pp.

Crepon, M. (1972). Generation of internal waves of inertia period in a two layer ocean. *Rapports et Procès-Verbaux, Intern. Council Expl. Sea (Copenhagen)* 162, 85-88.

Csanady, G. T. (1968). Motions in a model great lake due to a suddenly imposed wind. *J. Geophys. Res.* 73(20), 6435-6447.

Csanady, G. T. (1972). The coastal boundary layer in Lake Ontario. Part I. The spring regime. *J. Phys. Oceanogr.* 2(1), 41-53.

Darbyshire, J. (1970). The variation of the depth of a high salinity layer in the Celtic Sea. *Deep-Sea Res.* 17(5), 903-911.

Davis, R. E. (1969). The two-dimensional flow of a stratified fluid over an obstacle. *J. Fluid Mech.* 36(1), 127-143.

Davis, R. E. and A. Acrivos (1967a). Solitary internal waves in deep water. *J. Fluid Mech.* 29, 593-607.

Davis, R. E. and A. Acrivos (1967b). Stability of oscillatory internal waves. *J. Fluid Mech.* 30, 723-736.

Day, C. G. and F. Webster (1965). Some current measurements in the Sargasso Sea. *Deep-Sea Res.* 12, 805-814.

Deardorff, J. W., G. E. Willis, and D. K. Lilly (1969). Laboratory investigation of non-steady penetrative convection. *J. Fluid Mech.* 35(1), 7-31.

Defant, A. (1940a). Die ozeanographischen Verhältnisse während der Ankerstation des „Altair" am Nordran des Hauptstromstriches des Golfstromes nördlich der Azoren. *Ann. Hydrogr. u. Marit. Meteorol.* 68, 1-34.

Defant, A. (1940b). Scylla und Charybdis und die Gezeitenströmungen in der Strasse von Messina. *Ann. Hydrogr. u. Marit. Meteorol.* 68, 145-157.

Defant, A. (1961). Internal waves. In, *Physical oceanography,* Vol. 2. New York: Pergamon Press, pp. 517-570.

Defant, A. and O. V. Schubert (1934). Strommessungen und ozeanographische Serienbeobachtungen der Vier-Länder-Unternehmung im Kattegat, August 1931. *Veröff. Inst. Meeresk. Berlin, N.F.A.* 25, 3-60. (Cited by Krauss, 1966b).

DeLisi, D. P. and G. M. Corcos (1973). A study of internal waves in a wind tunnel. *Boundary-Layer Meteorology* 5, 121-137.

Demoll, R. (1922). Temperaturwellen (=Seiches) und Planktonwellen. *Arch. Hydrobiol.* 3, 313-320.

Derügin, K. M. (1933). Über die inneren Wellen im östlichen Teil des finnischen Golfes, pp. 1-8. In, Proc. 4th Hydrological Conference of the Baltic States, Leningrad, 1933, Vol. 3.

Dore, B. D. (1970). On progressive internal waves in non-uniform depths. *Pure Appl. Geophys.* 83(4), 87-97.

Dore, B. D. (1971). Internal wave motion in a non-homogeneous viscous fluid of variable depth. *Proc. Cambridge Phil. Soc.* 70, 157-167.

Dowling, G. B. (1966). "Low-Frequency Shallow-Water Internal Waves at Panama City, Florida." U.S. Navy Mine Defense Lab., Res. Rept. 313, 59 pp. (Reference supplied by L. H. Larsen, Dept. Oceanogr., Univ. Washington, Seattle.)

Drazin, P. G. (1969). Non-linear internal gravity waves in a slightly stratified atmosphere. *J. Fluid Mech.* 36(3), 433-446.

Drazin, P. G. (1970). Kelvin-Helmholtz instability of finite amplitude. *J. Fluid Mech.* 42(2), 321-335.

Düing, W. (1969). Observations with a pycnocline follower, pp. 11-27. In, 5th Ann. Conf. Marine Temp. Meas. Symposium, Miami, Florida, 18 June, 1969, sponsored by Marine Tech. Soc.

Düing, W. (1970). The pycnocline follower. *EOS Trans. Amer. Geophys. Un.* 51(4), 314 (abstract).

Düing, W. and D. Johnson (1972). High resolution current profiling in the Straits of Florida. *Deep-Sea Res.* 19, 259-274.

Dutton, J. A. and H. A. Panofsky (1970). Clear air turbulence: a mystery may be unfolding. *Science* 167(3920), 937-944.

Dyment, A. (1968). Sur les ondes de pesanteur permanents à l'approximation d'ordre 2 dans un écoulement plan sur fond horizontal de deux liquides non miscibles. *C.-R. Acad. Sci. Paris, Ser. A* 267, 839-841.

Eckart, C. (1960). *Hydrodynamics of oceans and atmospheres.* New York: Pergamon Press, 290 pp.

Eckart, C. (1961). Internal waves in the ocean. *Phys. Fluids* 4(7), 791-799.

Ekman, V. W. (1904). On dead-water. *Scientific Results of the Norwegian North Polar Expedition 1893-1896* 5(15), 1-152.

Ekman, V. W. (1053). Studies on ocean currents. Results of a cruise on board the "Armauer Hansen" in 1930 under the leadership of Björn Helland-Hansen. *Geofys. Publ.* 19(1), 122 pp.

Eliassen, A. and E. Palm (1961). On the transfer of energy in stationary mountain waves. *Geofys. Publ.* 22(3), 1-23.

El-Sabh, M. I., R. Glombitza, and O. M. Johannessen (1971). "On the Vertical Fluctuations of Hydrochemical Parameters in the Gulf of St. Lawrence, 1969." McGill Univ., Montreal, Canada. Marine Sciences Centre Ms. 19, 24 pp.

Emery, K. O. (1956). Deep standing internal waves in California basins. *Limnol. Oceanogr.* 1(1), 35-41.

Emery, K. O. and C. G. Gunnerson (1973). Internal swash and surf. *Proc. Natl. Acad. Sci.* 70, 2379-2380.

Eskinazi, S. (1968). *Principles of fluid mechanics.* Boston: Allyn and Bacon, Inc., 538 pp.

Etienne, A. (1970). Etude spectral des ondes internes et de la turbulence. *Cah. Océanogr.* 22, 657-685.

Ewing, G. (1950). Slicks, surface films and internal waves. *J. Mar. Res.* 9(3), 161-187.

Exner, F. M. (1908a). Ergebnisse einiger Temperaturregistrierungen im Wolfgansee. *S.-B. Akad. Wiss. Wien, Math.-Nat. Kl., Ser. IIa* 117, 1295-1315.

Exner, F. M. (1908b). Über eigentümliche Temperaturschwankungen von eintagiger Periode im Wolfgangsee. *S.-B. Akad. Wiss. Wien, Math.-Nat. Kl., Ser. IIa* 117, 9-26. (Cited by Krauss, 1966b.)

Exner, F. M. (1928). Über Temperaturseiches im Lunzer See. *Ann. Hydrogr. u. Marit. Meteorol.* 14, 14-20.

Faller, A. J. (1966). Sources of energy for the ocean circulation and a theory of the mixed layer, pp. 651-672. In, Proc. 5th U.S. Natl. Congr. Applied Mechanics, Univ. Minnesota.

Fedosenko, V. S. (1969). The influence of viscosity of internal waves of tsunami type. *Izv. Atmos. Ocean. Phys.* 5(12), 771-772.

Fedosenko, V. S. and L. V. Cherkesov (1968). Internal waves from submarine earthquakes. *Izv. Atmos. Ocean. Phys.* 4(11), 686-692.

Fjeldstad, J. E. (1933). Interne wellen. *Geofys. Publ.* 10(6), 1-35.

Fjeldstad, J. E. (1936). Results of tidal observations, Norwegian North Polar Expedition with the "Maud", 1918-1925, Scientific Results, Vol. 4, No. 4. Bergen: Geofysisk Institutt.

Fjeldstad, J. E. (1952). Observations of internal tidal waves. *Natl. Bur. Stand. Circular* 251, 39-45.

Fjeldstad, J. E. (1958). Ocean currents as an initial problem. *Geofys. Publ.* 20(7), 1-24.

Fjeldstad, J. E. (1964). Internal waves of tidal origin. Part 1. Theory and analysis of observations, pp. 1-73. Part 2. Tables, pp. 1-155. *Geofys. Publ.* 25(5).

Fofonoff, N. P. (1966). Internal Waves of Tidal Period. Unpubl. Manuscript. Wood Hole Oceanogr. Inst., Woods Hole, Mass, 8 pp.

Fofonoff, N. P. (1969). Spectral characteristics of internal waves in the ocean. *Deep-Sea Res.* 16(Suppl.) 58-71.

Frankignoul, C. J. (1969). "Note on Internal Waves in a Simple Thermocline Model." Woods Hole Oceanogr. Inst., WHOI Tech. Rept. 69-47, 38 pp.

Frankignoul, C. J. (1970a). Generation of transient nearly inertial internal waves by the interaction between internal waves and a geostrophic shear current. *Univ. Liège Cah. Méc. Math.* 26, 141-155.

Frankignoul, C. J. (1970b). The effect of weak shear and rotation on internal waves. *Tellus* 22, 194-203.

Frankignoul, C. J. (1972). Comments on the energy exchange between internal waves and a shear flow. *Mém. Soc. Roy. Sci. Liège* 2(6), 51-58.

Frankingoul, C. J. and E. J. Strait (1973). Correspondence and vertical propagation of the inertial-internal wave energy in the deep sea. *Mém. Soc. Roy. Sci. Liège* 4(6), 151-161.

Frassetto, R. (1960). A preliminary survey of thermal microstructure in the Strait of Gibraltar. *Deep-Sea Res.* 7, 152-162.

Gade, H. G. (1970). Internal waves, Vol. 2, pp. 92-95. References, Vol. 2, pp. 185-187. Figures, Vol. 3, pp. 18, 19, 49. In, Hydrographic investigations in the Oslofjord, a study of water circulation and exchange processes, Universitetet i Bergen.

Gade, H. G. and E. Ericksen (1969). "Notes on the Internal Tide and Shortperiodic Secondary Oscillations in the Strait of Gibraltar." Universitetet i Bergen, *Årbok Mat. Naturvit. Ser. 9*, 24 pp.

Gargett, A. E. (1968). Generation of internal waves in the Strait of Georgia. *Trans. Amer. Geophys. Un.* 49, 705 (abstract).

Gargett, A. E. (1970). Ph.D. thesis. "Internal Waves in the Strait of Georgia." Univ. British Columbia, Vancouver, B.C., 113 pp. (Reference supplied by P. H. LeBlond, Dept. Oceanogr., Univ. British Columbia.)

Gargett, A. E. and B. A. Hughes (1972). On the interaction of surface and internal waves. *J. Fluid Mech.* 52(1), 179-191.

Garrett, C. J. R. (1968). On the interaction between internal gravity waves and shear flow. *J. Fluid Mech.* 34, 711-720.

Garrett, C, J. R. and W. H. Munk (1971). Internal wave spectra in the presence of fine-structure. *J. Phys. Oceanogr.* 1, 196-202.

Garrett, C. J. R. and W. H. Munk (1972a). Oceanic mixing by breaking internal waves. *Deep-Sea Res.* 19(12), 823-832.

Garrett, C. J. R. and W. H. Munk (1972b). Space-time scales of internal waves. *Geophys. Fluid Dyn.* 3(3), 225-264.

Gaul, R. D. (1961a). Observations of internal waves near Hudson Canyon. *J. Geophys. Res.* 66(11), 3821-3830.

Gaul, R. D. (1961b). "The Occurrence and Velocity Distribution of Short-Term Internal Temperature Variations near Texas Tower No. 4." U.S. Navy Hydrographic Office, Washington, D.C., ASWEPS Rept. 1, 45 pp.

Gieskes, J. and K. Grasshoff (1969). A study in the variability in the hydrochemical factors in the Baltic Sea on the basis of two anchor stations September 1967 and May 1968. *Kiel. Meeresforsch.* 25(1), 105-132.

Glinskii, N. T. (1960). On vertical fluctuations of the water temperature in the Black Sea. *Bull. Izv. Acad. Sci. USSR, Geophys. Ser.* 452-458.

Glinskii, N. T. and S. G. Boguslavskii (1963). Influence of internal waves on vertical exchange in the ocean. *Bull. Izv. Acad. Sci. USSR, Geophys. Ser.* 10, 938-942.

Goldstein, S. (1931). On stability of superposed streams of fluids of different densities. *Proc. Roy. Soc. London, Ser. A* 132, 524-548.

Gonella, J. (1971). A local study of inertial oscillations in the upper layers of the ocean. *Deep-Sea Res.* 18, 775-788.

Gordon, D. and T. N. Stevenson (1972). Viscous effects in a vertically propagating internal wave. *J. Fluid Mech.* 56(4), 629-639, 2 plates.

Gossard, E. E. (1962). Vertical flux of energy into the lower ionosphere from internal gravity waves generated in the troposphere. *J. Geophys. Res.* 67(2), 745-757.

Gossard, E. E., D. R. Jensen, and J. H. Richter (1971). An analytical study of tropospheric structure as seen by high-resolution radar. *J. Atmos. Sci.* 28(5), 794-807.

Goulet, J. R., Jr. and B. J. Culverhouse, Jr. (1972). STD thermometer time constant. *J. Geophys. Res.* 77(24), 4588-4589.

Greenhill, A. G. (1887). Wave motion in hydrodynamics. *Amer. J. Math.* 9, 62-112.

Greenspan, H. P. (1968). *The theory of rotating fluids*. London: Cambridge Univ. Press, 327 pp.

Gregg, M. C. (1973). The microstructure of the ocean. *Sci. Am.* 228(2), 64-77.

Grimshaw, R. H. J. (1968). A note on steady two-dimensional flow of a stratified fluid over an obstacle. *J. Fluid Mech.* 33, 293-301.

Grimshaw, R. H. J. (1971). Nonlinear internal gravity waves in a slowly varying medium. *J. Fluid Mech.* 54, 193-207.

Grimshaw, R. H. J. (1973). "Internal Gravity Waves in a Slowly Varying, Dissipative Medium." Dept. Math., Univ. Melbourne, Victoria, Australia. App. Math. Preprint 11, 26 pp.

Groen, P. (1948a). Contributions to the theory of internal waves. *Konin. Neder. Met. Inst. Bilt. Med. en Verhand., Ser. B* 2(11), 23 pp.

Groen, P. (1948b). Two fundamental theorems on gravity waves in inhomogeneous incompressible fluids. *Physica* 14, 294-300.

Gustafson, J. and B. Kullenberg (1936). Untersuchungen von Trägheitsströmungen in der Ostsee. *Svenska Hydrogr.-Biol. Komm. Skr. Ser.-Hydrografi* 13, 28 pp.

Hachmeister, L. E. and S. Martin (1973). Observations of resonant internal wave interactions over a range of driving frequencies. *EOS Trans. Amer. Geophys. Un.* 54(4), 308 (abstract).

Halbfass, W. (1909). Zur Frage der Temperaturseiches. *Petermanns Geographische Mitteilungen* 55, 364-365.

Halbfass, W. (1911). Gibt es im Madüsee Temperaturseiches? *Intern. Rev. Ges. Hydrobiol.-Hydrog.* 3, 1-40.

Hale, A. M. (1965). Internal waves in Lake Huron, pp. 47-59. In, "Mixing Processes and Internal Waves, Baie du Doré Report 1965," Great Lakes Inst., Univ. Toronto, Rept. PR 26.

Hale, A. M. (1969). Internal waves of the second vertical mode in Lake Huron, pp. 564-566. In, Proc. 12th Conf. Great Lakes Res., sponsored by Intern. Assoc. Great Lakes Res.

Hall, M. J. and Y.-H. Pao (1969). Spectra of internal waves and turbulence in stratified fluids. Part 2. Experiments on the breaking of internal waves in a two-fluid system. *Radio Sci.* 4(12), 1321-1325.

Hall, M. J. and Y.-H. Pao (1971). "Internal Wave Breaking in a Two-Fluid system." Boeing Sci. Res. Lab., Seattle, Wash., Doc. D1-82-1076, 141 pp.

Halpern, D. (1971a). Observations on short-period internal waves in Massachusetts Bay. *J. Mar. Res.* 29(2), 116-132.

Halpern, D. (1971b). Semidiurnal internal tides in Massachusetts Bay. *J. Geophys. Res.* 76(27), 6573-6584.

Harrison, W. J. (1908). The influence of viscosity on the oscillations of superposed fluids. *Proc. London Math. Soc., Ser. 2* 6, 396-405.

Hartman, R. J. and H. W. Lewis (1972). Wake collapse in a stratified fluid: linear treatment. *J. Fluid Mech.* 51(3), 613-618.

Hasselmann, K. (1966). Feynman diagrams and interaction rules of wave-wave scattering processes. *Rev. Geophys.* 4(1), 1-32.

Hasselmann, K. (1967). A criterion for nonlinear wave stability. *J. Fluid Mech.* 30(4), 737-739.

Hasselmann, K. (1970). Wave-driven inertial oscillations. *Geophys. Fluid Dyn.* 1, 463-502.

Haurwitz, B. (1948). The effect of ocean currents on internal waves. *J. Mar. Res.* 7, 217-228.

Haurwitz, B. (1954). The occurrence of internal tides in the ocean. *Arch. Met. Geophys. Bioklimatol., Ser. A* 7, 406-424.

Haurwitz, B., H. Stommel, and W. H. Munk (1959). On the thermal unrest in the ocean. In, *The atmosphere and the sea in motion*, B. Bolin (Ed.). New York: Rockefeller Inst. Press, pp. 74-94.

Hazel, P. (1967). The effect of viscosity and heat conduction on internal gravity waves at a critical layer. *J. Fluid Mech.* 30, 775-783.

Hazel, P. (1972). Numerical studies of the stability of inviscid stratified shear flows. *J. Fluid Mech.* 51(1), 39-61.

Heaps, N. S. and A. E. Ramsbottom (1966). Wind effects on the water in a narrow two-layered lake. *Phil. Trans. Roy. Soc. London, Ser. A* 259(1102), 391-430.

Hecht, A. and P. Hughes (1971). Observations of temperature fluctuations in the upper layers of the Bay of Biscay. *Deep-Sea Res.* 18(7), 663-684.

Hecht, A. and R. A. White (1968). Temperature fluctuations in the upper layer of the ocean. *Deep-Sea Res.* 15, 339-353.

Hela, I. and W. Krauss (1959). Zum Problem der starken Veränderlichkeit des Schichtungsverhältnisse im Arkona-Becken. *Kiel. Meeresforsch.* 15(2), 125-143.

Helland-Hansen, B. (1930). Physical oceanography and meteorology. In, *Rep. Scient. Res. "Michael Sars" North Atlant. Deep Sea Exp. 1910*, Vol. 1. Bergen: J. Griegs, pp. 1-115. (Cited by Fjeldstad, 1964).

Helland-Hansen, B. and F. Nansen (1926). The eastern North Atlantic. *Geofys. Publ.* 4(2), 76 pp., 71 plates.

Hendershott, M. C. (1968). "Inertial Oscillations of Tidal Period." Scripps Inst. Oceanogr. Ref. 68-12, 135 pp.

Hendershott, M. C. (1969). Impulsively started oscillations in a rotating stratified fluid. *J. Fluid Mech.* 36, 513-527.

Hendershott, M. C. (1973). Ocean tides. *EOS Trans. Amer. Geophys. Un.* 54(2), 76-86.

Hines, C. O. (1968). A possible source of waves in noctilucent clouds. *J. Atmos. Sci.* 25(5), 937-942.

Hinwood, J. B. (1972). The study of density-stratified flows up to 1945. Part 2. Internal waves and interfacial effects. *La Houille Blanche* 8, 709-722.

Hogg, N. G. (1971). Longshore current generation by obliquely incident internal waves. *Geophys. Fluid Dyn.* 2(4), 361-376.

Hollan, E. (1966a), Das Spektrum der internen Bewegungsvorgänge der westlichen Ostsee im Periodenbereich von 0,3 bis 60 Minuten. Teil 1. Interpretation der Wellen förmigen Bewegungsanteile. *Dtsche. Hydrogr. Zeit.* 19(6), 193-218.

Hollan, E. (1966b). Das Spektrum der internen Bewegungsvorgänge der westlichen Ostsee im Periodenbereich von 0,3 bis 60 Minuten. Teil 2. Beobachtungsergebnisse. *Dtsch. Hydrogr. Zeit.* 19(6), 285-298.

Hollan, E. (1969). Die Veränderlichkeit der Strömungsverteilung im Gotland-Becken am Beispeil von Strömungsmessungen im Gotland-Tief. *Kiel. Meeresforsch.* 25(1), 19-70.

Hollan, E. (1972). Kurzperiodische interne Wellen des Meeres als Folge Kleinräumiger kurzzeitig wirksamer äusserer Kräfte. *Dtsch. Hydrogr. Zeit.* 25(1), 14-37.

Hooke, W. H. (1973). Further study of the "jet-stream-associated" atmospheric gravity waves over the eastern seabord on March 18, 1969. *EOS Trans. Amer. Geophys. Un.* 54(4), 290 (abstract).

Horn, W., W. Hussels, and J. Meincke (1971). Schichtungs- und Strömungsmessungen im Bereich der Grossen Meteorbank. *„Meteor" Forsch.-Ergebnisse A* 9, 31-46.

Howard, L. N. (1961). Note on a paper of John W. Miles. *J. Fluid Mech.* 10, 509-512.

Hudimac, A. A. (1961). Ship waves in a stratified ocean. *J. Fluid Mech.* 11, 229-243.

Hughes, B. A. (1969). "Characteristics of Some Internal Waves in Georgia Strait." Defence Res. Est. Pacific, Victoria, British Columbia, Canada, Tech. Memo. 69-2, 9 pp. 17 figs.

Hunkins, K. and M. Fliegel (1973). Internal undular surges in Seneca Lake; a natural occurrence of solitons. *J. Geophys. Res.* 78(3), 539-548.

Hunt, J. N. (1961). Interfacial waves of finite amplitude. *La Houille Blanche* 4, 515-525.

Hurley, D. G. (1969). The emission of internal waves by vibrating cylinders. *J. Fluid Mech.* 36(4), 657-672.

Hurley, D. G. (1970). Internal waves in a wedge-shaped region. *J. Fluid Mech.* 43(1), 97-120.

Hurley, D. G. (1972). A general method for solving steady-state internal gravity wave problems. *J. Fluid Mech.* 56(4), 721-740.

Ichiye, T. (1963). "Internal Waves Over a Continental Shelf." Oceanogr. Inst. Florida State Univ., Tech. Rept. 30, 23 pp.

Ivanov, Yu, A., B. A. Smirnov, B. A. Tareev, and B. N. Filyushkin (1969). Experimental investigation of temperature fluctuations in the sea in the range of frequencies of internal gravity waves. *Izv. Atmos. Ocean. Phys.* 5(4), 230-235.

Iwata, N. (1962). Effekt der Meeresströmung auf interne Wellen im offenen Meer. *J. Oceanogr. Soc. Japan* 18(2), 69-72.

Jacobsen, J. P. and H. Thomsen (1934). Periodic variations in temperature and salinity in the Straits of Gibraltar. In, *James Johnstone memorial volume*, Liverpool: Liverpool Univ. Press, pp. 275-293.

Janowitz, G. S. (1968). On wakes in stratified fluids. *J. Fluid Mech.* 33(3), 417-432.

Jenkins, G. M. and D. G. Watts (1968). *Spectral analysis and its applications*. San Francisco: Holden-Day, 525 pp.

Johannessen, O. M. (1968). Some current measurements in the Drøback Sound, the narrow entrance to the Oslofjord. *Hvalrådets Skrifter* 50, 38 pp.

Johns, B. and M. J. Cross (1969). Decay of internal wave modes in a multi-layered system. *Deep-Sea Res.* 16, 185-195.

Johns, B. and M. J. Cross (1970). The decay and stability of internal wave modes in a multi-sheeted thermocline. *J. Mar. Res.* 28(2), 215-224.

Jones, W. L. (1967). Propagation of internal gravity waves in fluids with shear flow and rotation. *J. Fluid Mech.* 30, 439-448.

Jones, W. L. (1968). Reflexion and stability of waves in stably stratified fluids with shear flow: a numerical study. *J. Fluid Mech.* 34, 609-624.

Jones, W. L. (1969a). Ray tracing for internal gravity waves. *J. Geophys. Res.* 74(8), 2028-2033.

Jones, W. L. (1969b). The transport of energy by internal waves. *Tellus* 21(2), 177-184.

Joyce, T. M. (1972). "Nonlinear Interactions among Standing Surface and Internal Gravity Waves." Woods Hole Oceanogr. Inst, WHOI Rept. 72-3, 96 pp.

Joyce, T. M. (1973). Fine structure contamination of measured internal waves: a numerical experiment. *EOS Trans. Amer. Geophys. Un.* 54(4), 322 (abstract).

Kahalas, S. L. and B. L. Murphy (1971). Acoustic-gravity wave generation at 120 km altitude by sea level detonation: a preliminary analysis of the Greene-Whitaker calculation. *Geophys. J. Roy. Astron. Soc.* 26(1-4), 391-398.

Kanari, S. (1968). On the studies of internal waves in Lake Biwa (in Japanese, with English synopsis and figure labels). *Annals, Disaster Prevention Res. Inst., Kyoto Univ., Japan* 11B, 179-189.

Kanari, S. (1970a). Internal waves in Lake Biwa (I). The responses of the thermocline to the wind action (in English). *Bulletin, Disaster Prevention Res. Inst., Kyoto Univ. Japan* 19, 19-26.

Kanari, S. (1970b). On the studies of internal waves in Lake Biwa (III). On the measurement of the vertical displacement of waters by using an instrumented neutrally-buoyant float (in Japanese, with English synopsis and figure labels). *Annals, Disaster Prevention Res. Inst., Kyoto Univ., Japan* 13A, 601-608.

Kanari, S. (1973). Internal waves in Lake Biwa (II). Numerical experiments with a two layer model. *Bulletin, Disaster Prevention Res. Inst. Kyoto Univ., Japan* 22, Part 2(202), 69-96.

Käse, R. H. (1971). Über zweidimensionale luftdruckbedingte interne Wellen im exponentiell geschichteten Meer. *Dtsch. Hydrogr. Zeit.* 24(5), 193-209.

Kaylor, R. and A. J. Faller (1972). Instability of the stratified Ekman boundary layer and generation of internal waves. *J. Atmos. Sci.* 29(3), 497-509.

Keady, G. (1971). Upstream influence in a two-fluid system. *J. Fluid Mech.* 49(2), 373-384.

Keller, J. B. and V. C. Mow (1969). Internal wave propagation in an inhomogeneous fluid of nonuniform depth. *J. Fluid Mech.* 38, 365-674.

Keller, J. B. and W. H. Munk (1970). Internal wave wakes of a body moving in a stratified fluid. *Phys. Fluids* 13, 1425-1431.

Kelly, R. E. (1969). Wave diffraction in a two-fluid system. *J. Fluid Mech.* 36(1), 65-73.

Kelly, R. E. (1970). Wave induced boundary layers in a stratified fluid. *J. Fluid Mech.* 42(1), 139-150.

Kelly, R. E. and S. A. Maslowe (1970). The nonlinear critical layer in a slightly stratified shear flow. *Stud. Appl. Math.* 49(4), 301-326.

Kelly, R. E. and J. D. Vreeman (1970). Excitation of waves and mean currents in a stratified fluid due to a moving heat source. *Zeit. Angew. Math. Phys.* 21(1), 1-16.

Kenyon, K. E. (1968). Wave-wave interactions of surface and internal waves. *J. Mar. Res.* 26(3), 208-231.

Keulegan, G. H. (1953). Characteristics of internal solitary waves. *J. Res. Nat. Bur. Stand.* 51, 133-140.

Keulegan, G. H. and L. H. Carpenter (1961). "An Experimental Study of Internal Progressive Oscillatory Waves." Nat. Bur. Stand. Rept. 7319, 34 pp.

Keunecke, K.-H. (1970). Stehende interne Wellen in rechteckigen Becken. *Dtsch. Hydrogr. Zeit.* 23, 61-79.

Keunecke, K.-H. (1971a). "A filter Technique for Repeated Hydrographic Sections." Forschung. der Bundeswehr für Wasserschall- und Geophysik, Kiel, Germany, FWG-Bericht 1971-3, 19 pp.

Keunecke, K.-H. (1971b). "Interne Gezeiten am Kontinentalabhang während des Expedition „Norwegische See 1969." Forschung. der Bundeswehr für Wasserschall- und Geophysik, Kiel, Germany, FWG Bericht 1971-7, 7 pp.

Keunecke, K.-H. (1972). On the observation of internal tides at the continental slope of Norway. *EOS Trans. Amer. Geophys. Un.* 53(4), 396 (abstract).

Kielman, J., W. Krauss, and L. Magaard (1969). Über die Verteilung der kinetischen Energie im Bereich der Trägheits- und Seiches-frequenzen der Ostsee im August 1964. *Kiel. Meeresforsch.* 25(2), 245-254.

Kinsman, B. (1965). Capillary and internal small-amplitude waves. In, *Wind waves*, Englewood Cliffs, N. J.: Prentice-Hall, pp. 167-174.

Knauss, J. A. (1962). Observations of internal waves of tidal period made with neutrally buoyant floats. *J. Mar. Res.* 20(2), 111-118.

Konaga, S. (1961). Kuroshio in Enshū-nada (III): Variations of the depth of isotherms. *Oceanogr. Mag. Tokyo* 13(1), 31-40.

Konaga, S. (1965). Observation of the internal waves. *Oceanogr. Mag. Tokyo* 17, 141-174.

Krauss, W. (1959). Über meteorologisch bedingte interne Wellen auf einer Dauerstation südwestlich Islands. *Dtsch. Hydrogr. Zeit., Ergänzungsheft Reihe B (4°)* No. 3, 55-58.

Krauss, W. (1961). Meteorologically forced internal waves in the region south-west of Iceland. *Rapports et Procès-Verbaux, Intern. Council Expl. Sea (Copenhagen)* 149, 89-92.

Krauss, W. (1963). Zum System der internen Seiches des Ostsee. *Kiel. Meeresforsch.* 19, 119-132. (Cited by Krauss, 1966b).

Krauss, W. (1966a). "Internal Tides off Southern California--Analysis of Data Obtained by NEL Thermistor Chain." Navy Electronics Lab. Rept. 1389, 53 pp.

Krauss, W. (1966b). *Methoden und Ergebnisse der theoretischen Ozeanographie, B. II--Interne Wellen.* Berlin: Gebrüder Borntraeger, 248 pp.

Krauss, W. (1967). "Interaction between Surface and Internal Waves in Shallow Water;" Navy Electronics Lab. Rept. 1432, 28 pp.

Krauss, W. (1969). Typical features of internal wave spectra. In, *Progress in oceanography*, Vol. 5. M. Sears (Ed.), New York: Pergamon Press, pp. 95-103.

Krauss, W. (1972). On the response of a stratified ocean to wind and air pressure. *Dtsch. Hydrograph. Zeit.* 25(2), 49-61.

Krauss, W. and W. Düing (1963). Schichtung und Strom im Golf von Neapel. *Publ. Stazione Zoologica Napoli* 33, 243-263.

Krauss, W. and L. Magaard (1961). Zum Spektrum der internen Wellen der Ostsee. *Kiel. Meeresforsch.* 17, 137-147.

Kullenberg, B. (1935). Interne Wellen im Kattegat. *Svenska Hydrogr.-Biol. Komm. Skr. Ser.-Hydrografi.* 12, 17 pp. (Translated by Great Britain Admiralty. Dept. of Res. Programmes and Planning, ACSIL Transl. 561, September 1952, Internal waves in the Kattegat.)

Lacombe, H. (1965). Courants de densité dans le détroit de Gibraltar. *La Houille Blanche* 1, 38-44.

LaFond, E. C. (1959). Sea surface features and internal waves in the sea. *Indian J. Meteorol. Geophys.* 10, 415-419.

LaFond, E. C. (1961). Boundary effects on the shape of internal temperature waves. *Indian J. Meteorol. Geophys.* 12, 335-338.

LaFond, E. C. (1962). Internal waves, Part 1. In, *The sea*, Vol. 1, M. N. Hill (Ed.). New York: Interscience, pp. 731-751.

LaFond, E. C. (1963). Detailed temperature structures of the sea off Baja California. *Limnol. Oceanogr.* 8(4), 417-425.

LaFond, E. C. (1966). Internal waves. In, *Encyclopedia of oceanography*, New York: Reinhold Publ., pp. 402-408.

LaFond, E. C. and K. G. LaFond (1967). Internal thermal structures in the ocean. *J. Hydronautics* 1(1), 48-53.

LaFond, E. C. and A. T. Moore (1962). "Measurements of Thermal Structure off Southern California with the NEL Thermistor Chain." Navy Electronics Lab. Rept. 1130 (Cited by Krauss, 1966b.)

LaFond, E. C. and C. P. Rao (1954). Vertical oscillations of tidal periods in the temperature structure of the sea. *Andhra Univ. Mem. Oceanogr.* 1, 109-116.

Lahti, B. P. (1968). *An introduction to random signals and communication theory.* Scranton, PA: International Textbook Co., 488 pp.

Lamb, H. (1916). On waves due to a travelling disturbance, with an application to waves in superposed fluids. *Phil. Mag.* 31(6), 386-399.

Lamb, H. (1945). Articles 231-235. In, *Hydrodynamics*, 6th ed., New York: Dover Publ. pp. 370-380.

Larsen, L. H. (1969a). Internal waves and their role in mixing the upper layers of the ocean. *The Trend in Engineering, Univ. of Washington, Seattle, Washington* 21(4), pp. 8-11 and p. 19.

Larsen, L. H. (1969b). Internal waves incident upon a knife edge barrier. *Deep-Sea Res.* 16, 411-419.

Larsen, L. H. (1969c). Oscillations of a neutrally buoyant sphere in a stratified fluid. *Deep-Sea Res.* 16, 587-603.

Larsen, L. H., M. Rattray, Jr., W. Barbee, and J. G. Dworski (1972). Internal tides. *Rapports et Procès-Verbaux, Intern. Council Expl. Sea (Copenhagen)* 162, 65-79.

Laykhtman, D. L., A. I. Leonov, and Yu. Z. Miropol'skii (1971). Interpretation of measurements of statistical parameters of scalar fields in the ocean in the presence of internal gravity waves. *Izv. Atmos. Ocean. Phys.* 7(4), 291-295.

Lazier, J. R. N. (1973). T-S structures interpreted as effects of long period internal waves. *EOS Trans. Amer. Geophys. Un.* 54(4), 322 (abstract).

LeBlond, P. H. (1966). On the damping of internal gravity waves in a continuously stratified ocean. *J. Fluid Mech.* 25(1), 121-142.

LeBlond, P. H. (1970). Internal waves in a fluid of finite Prandtl number. *Geophys. Fluid Dyn.* 1, 371-376.

Lee, C.-Y. (1972). Ph.D. thesis. "Long Nonlinear Internal Waves and Quasi-Steady Lee Waves." MIT, and Woods Hole Oceanogr. Inst., WHOI Rept. 72-4, 127 pp.

Lee, O. S. (1961). Observations of internal waves in shallow water. *Limnol. Oceanogr.* 6(3), 312-321.

Lee, W. H. K. and C. S. Cox (1966). Time variation of ocean temperatures and its relation to internal waves and oceanic heat flow measurements. *J. Geophys. Res.* 71(8), 2101-2111.

Lee, Y. W. (1960). *Statistical theory of communication.* New York: J. Wiley and Sons, Inc., 509 pp.

Le Floc'h, J. (1970). Sur quelques observations de fluctuations de temperature et de vitesse de courant associées à des ondes internes à courte periode ou à la turbulence. *Cah. Océanogr.* 22(7), 687-699.

Lek, D. L. (1938). Interne Wellen in den niederlandischostindischen Gewässern. *Comptes-Rendus, 12th Intern. Geographic Congress* 12(2b), 69-76.

Lek, D. L. and J. E. Fjeldstad (1938). Die Ergebnisse der Strom- und Serienmessungen--Interne Wellen auf Ankerstation 253A. *The Snellius Exp. Oceanogr. Res.* 2, 152-169.

Lighthill, M. J. (1967). On waves generated in dispersive systems by travelling forcing effects, with applications to the dynamics of rotating fluids. *J. Fluid Mech.* 27(4), 725-752.

Lilly, D. K. (1972). Wave momentum flux--a GARP problem. *Bull. Amer. Met. Soc.* 53(1), 17-23.

Lisitzin, E. (1953). On internal waves in the Northern Baltic. *Havsforsknings Inst. Skr.* 161, 1-9.

Liu, C. H. (1970). Progagation of acoustic gravity waves in a turbulent atmosphere. *Ann. Geophys.* 26, 35-41.

Lobb, G. and G. R. Hamilton (1969). Detection of internal waves from deep thermistor tows. *EOS Trans. Amer. Geophys. Un.* 50(4), 188 (abstract).

Lofquist, K. (1970). "Internal Waves Produced by Spheres Moving in Density Stratified Water." Nat. Bur. Stand. Rept. 10267, 51 pp.

Long, R. R. (1953). Some aspects of the flow of stratified fluids. I. A theoretical investigation. *Tellus* 5, 42-58.

Long, R. R. (1954a). Some aspects of the flow of stratified fluids. II. Experiments with a two-fluid system. *Tellus* 6, 97-115.

Long, R. R. (1954b). Some aspects of the flow of stratified fluids. III. Continuous density gradients. *Tellus* 7, 341-347.

Long, R. R. (1956a). Long waves in a two-fluid system. *J. Meteorol.* 13, 70-74.

Long, R. R. (1956b). Solitary waves in the one and two fluid systems. *Tellus* 8, 460-471.

Long, R. R. (1965). On the Boussinesq approximation and its role in the theory of internal waves. *Tellus* 17, 46-52.

Long, R. R. (1972). The steepening of long, internal waves. *Tellus* 24(2), 88-99.

Long, R. R. and J. B. Morton (1966). Solitary waves in compressible stratified fluids. *Tellus* 18, 79-85.

Longuet-Higgins, M. S. (1969). On the reflexion of wave characteristics from rough surfaces. *J. Fluid Mech.* 37(2), 231-250.

Loucks, R. H. (1963). An estimate of the scale-length of internal waves in the seasonal thermocline near Ocean Station P. *Canada Fish. Res. Bd., Manuscript Rept. Ser. (Oceanogr. Limn.)* 171, 1-41.

McEwan, A. D. (1971). Degeneration of resonantly-excited standing internal gravity waves. *J. Fluid Mech.* 50(3), 431-448.

McEwan, A. D. (1973). Interactions between internal gravity waves and their traumatic effect on a continuous stratification. *Boundary-Layer Meteorology* 5, 159-175.

McEwan, A. D., D. W. Mander and R. K. Smith (1972). Forced resonant second-order interaction between damped internal waves. *J. Fluid Mech.* 55(4), 589-608, 2 plates.

McGorman, R. E. (1972). Master's thesis. "Internal Waves in a Randomly Stratified Ocean." Univ. British Columbia, Vancouver, B.C., 56 pp.

McGorman, R. E. and L. A. Mysak (1973). Internal waves in a randomly stratified fluid. *Geophys. Fluid Dyn.* 4, 243-266.

McIntyre, M. E. (1972). On Long's hypothesis of no upstream influence in uniformly stratified or rotating flow. *J. Fluid Mech.* 52(2), 209-243.

McIntyre, M. E. (1973). Mean motions and impulse of a guided internal gravity wave packet. *J. Fluid Mech.* 60(4), 801-811.

McLaren, T. I., A. D. Pierce, T. Fohl, and B. L. Murphy (1973). An investigation of internal gravity waves generated by a buoyantly rising fluid in a stratified medium. *J. Fluid Mech.* 57(2), 229-240, 1 plate.

McWilliams, J. (1972). Observations of kinetic energy correspondences in the internal wave field. *Deep-Sea Res.* 19(11), 793-811.

Maeda, A. (1971). Phase velocity of semi-diurnal internal waves at Ocean Weather Station T. *J. Oceanogr. Soc. Japan* 27(4), 163-174.

Magaard, L. (1962). Zur Berechnung interner Wellen im Meeresräumen mit nicht-ebenen Böden bei einer speziellen Dichtverteilung. *Kiel. Meeresforsch.* 18, 161-183.

Magaard, L. (1965). Zur Theorie zweidimensionaler nichtlinearer interner Wellen in stetig geschichteten Medien. *Kiel. Meeresforsch.* 21(1), 22-32.

Magaard, L. (1968). Ein Beitrag zur Theorie der internen Wellen als Störungen geostrophischer Strömungen. *Dtsch. Hydrogr. Zeit.* 21(2), 241-278.

Magaard, L. (1971). Zur Berechnung von 'uftdruck- und windbedingten bewegungen eines stetig geschichteten seitlich unbegrenzten Meeres. *Dtsch. Hydrogr. Zeit.* 24(4), 145-158.

Magaard, L. (1973). On the generation of internal gravity waves by a fluctuating buoyancy flux at the sea surface. *Geophys. Fluid Dyn.* 5, 101-111.

Magaard, L. and W. Krauss (1967). Internal waves at Diamond Stations during the International Iceland-Färoe Ridge Expedition, May-June, 1960. *Rapports et Procès-Verbaux, Intern. Council Expl. Sea (Copenhagen)* 157, 173-183.

Magaard, L. and W. D. McKee (1973). Semi-diurnal tidal currents at 'Site D'. *Deep-Sea Res.* 20, 997-1009.

Magnitzky, A. W. and H. V. French (1960). "Tongue of the Ocean Research Experiment." Hydrographic Office, Washington, D.C., Tech. Rept. 94, 132 pp.

Malkus, J. S. and M. E. Stern (1953). The flow of a stable atmosphere over a heated island. Part 1. *J. Meteorol.* 10, 30-41.

Malone, F. D. (1968). An analysis of current measurements in Lake Michigan. *J. Geophys. Res.* 73(22), 7065-7081.

Manton, M. J., L. A. Mysak, and R. E. McGorman (1970). The diffraction of internal waves by a semi-infinite barrier. *J. Fluid Mech.* 43(1), 165-176.

Marchuk, G. I. and B. A. Kagan (1970). Internal gravity waves in a real stratified ocean. *Izv. Atmos. Ocean. Phys.* 6, 236-241.

Martin, S., W. F. Simmons, and C. I. Wunsch (1969). Resonant internal wave interactions. *Nature* 224, 1014-1016.

Martin, S., W. F. Simmons, and C. I. Wunsch (1972). The excitation of resonant triads by single internal waves. *J. Fluid Mech.* 53(1), 17-44.

Maslowe, S. A. and R. E. Kelly (1970). Finite-amplitude oscillations in a Kelvin-Helmholtz flow. *Intern. J. Non-linear Mech.* 5, 427-435.

Maslowe, S. A. and R. E. Kelly (1971). Inviscid instability of an unbounded heterogeneous shear layer. *J. Fluid Mech.* 48(2), 405-415.

Mei, C. C. (1968). Collapse of a homogeneous fluid mass in a stratified fluid, pp. 321-330. In, Proc. 12th International Congress of Applied Mech., Stanford, Aug. 1968.

Mei, C. C. and T. Y.-T. Wu (1964). Gravity waves due to a point disturbance in a plane free surface flow of stratified fluids. *Phys. Fluids* 7, 1117-1133.

Meincke, J. (1971a). Der Einfluss der Grossen Meteorbank auf Schichtung und Zirkulation der ozeanischen Deckschicht. *„Meteor" Forsch.-Ergebnisse A* 9, 67-94.

Meincke, J. (1971b). Observation of an anticyclonic vortex trapped above a seamount. *J. Geophys. Res.* 76(30), 7432-7440.

Milder, M. (1973). "User's Manuel for the Computer Program ZMODE." R and D Associates, Santa Monica, Calif., RDA-TR-2701-001, 55 pp.

Miles, J. W. (1961). On the stability of heterogeneous shear flows. *J. Fluid Mech.* 10, 496-508.

Miles, J. W. (1963). On the stability of heterogeneous shear flows. Part II. *J. Fluid Mech.* 16, 209-227.

Miles, J. W. (1967). Internal waves in a continuously stratified atmosphere or ocean. *J. Fluid Mech.* 28(2), 305-310.

Miles, J. W. (1968). Waves and wave drag in stratified flows, pp. 50-76. In, Proc. 12th International Congress of Applied Mech., Stanford, Aug. 1968.

Miles, J. W. (1971). Internal waves generated by a horizontally moving source. *Geophys. Fluid Dyn.* 2, 63-87.

Miles, J. W. (1972). Internal waves in a sheeted thermocline. *J. Fluid Mech.* 53(3), 557-573.

Miles, J. W. and H. E. Huppert (1969). Lee waves in a stratified flow. Part 4. Perturbation approximations. *J. Fluid Mech.* 35(3), 497-525.

Miropol'skii, Yu. Z. (1972). The effect of the microstructure of the density field in the sea on the propagation of internal gravity waves. *Izv. Atmos. Ocean. Phys.* 8(8), 515-517.

Mooers, C. N. K. (1970). The effects of horizontal density gradients and sloping boundaries on the propagation of inertial-internal waves. *Univ. Liège Cah. Méc. Math.* 26, 108-140.

Mooers, C. N. K. (1972). The mixed initial-boundary value problem for inertial-internal waves in a wedge. *Rapports et Procès-Verbaux, Intern. Council Expl. Sea (Copenhagen)* 162, 57-64.

Mooers, C. N. K. Seiche conditions for internal tides in the Florida Current. Unpubl. manuscript. (Reference supplied by C. N. K. Mooers, RSMAS, Univ. Miami, Florida.)

Mork, M. (1968a). "On the Formation of Internal Waves Caused by Tidal Flow Over a Bottom Irregularity." Univ. Bergen Geof. Inst., 28 pp.

Mork, M. (1968b). "The response of a Stratified Sea to Atmospheric Forces." Univ. Bergen Goef. Inst., 41 pp.

Mork, M. and H. G. Gade (1967). Internal waves. In, *International dictionary of geophysics*, New York: Pergamon Press, pp. 746-755.

Morse, B. A. and J. D. Smith (1972). Density structure of the mixed layer under multi-year ice. *EOS Trans. Amer. Geophys. Un.* 53(11), 1011 (abstract).

Mortimer, C. H. (1951). The use of models in the study of water movements in stratified lakes. *Proc. Intern. Assoc. Theoret. Appl. Limnol.* 11, 254-260.

Mortimer, C. H. (1952). Water movements in lakes during summer stratification; evidence from the distribution of temperature in Windermere. *Phil. Trans. Roy. Soc., London, Ser. B.* 236, 355-404.

Mortimer, C. H. (1953). The resonant response of stratified lakes to wind. *Schweiz. Zeit. Hydrol.* 15, 94-151.

Mortimer, C. H. (1963). Frontiers in physical limnology with particular reference to long waves in rotating basins, pp. 9-42. In, Proc. 6th Conf. Great Lakes Research, Great Lakes Research Division, Univ. Mich. Pub. 10.

Mortimer, C. H. (1968). "Internal Waves and Associated Currents Observed in Lake Michigan During the Summer of 1963." Center for Great Lakes Studies, Univ. Wisconsin, Milwaukee, Special Rept. 1, 24 pp., 120 figs.

Mortimer, C. H. (1971). "Large-Scale Oscillatory Motions and Seasonal Temperature Changes in Lake Michigan and Lake Ontario." Part 1, Text, 111 pp., Part II, Illustrations, 106 pp. Center for Great Lakes Studies, Univ. Wisconsin, Milwaukee, Special Rept. 12.

Mowbray, D. E. and B. S. H. Rarity (1967a). A theoretical and experimental investigation of the phase configuration of internal waves of small amplitude in a density stratified fluid. *J. Fluid Mech.* 28, 1-16.

Mowbray, D. E. and B. S. H. Rarity (1967b). The internal wave pattern produces by a sphere moving vertically in a density stratified fluid. *J. Fluid Mech.* 30(3), 489-495.

Munk, W. H. (1941). Internal waves in the Gulf of California. *J. Mar. Res.* 4(1), 81-91.

Munk, W. H. and C. J. R. Garrett (1973). Internal wave breaking and microstructure. *Boundary-Layer Meteorol.* 4, 37-45.

Munk, W. H. and N. Phillips (1968). Coherence and band structure of inertial motion in the sea. *Rev. Geophys.* 6(4), 447-472.

Mysak, L. A. (1973). The effect of a randomly perturbed Väisälä frequency on the propagation properties of internal waves. *Mém. Soc. Roy. Sci. Liège* 6(4), 171-181.

Nagashima, H. (1971). Reflection and breaking of internal waves on a sloping beach. *J. Oceanogr. Soc. Japan* 27(1), 1-6.

Nan'niti, T. and M. Yasui (1957). Fluctuations of water temperature off Turumi, Yokahama (in Japanese, with English abstract and figure titles). *J. Oceanogr. Soc. Japan* 13, 1-2.

Neshyba, S. J., B. T. Neal, and W. W. Denner (1972). Spectra of internal waves: *In situ* measurements in a multiple-layered structure. *J. Phys. Oceanogr.* 2, 91-95.

Nesterov, S. V. (1970). Resonance generation of internal waves. *Izv. Atmos. Ocean. Phys.* 6, 437-438.

Neumann, G. and W. J. Pierson, Jr. (1966). *Principles of physical oceanography.* Englewood Cliffs, N.J.: Prentice-Hall, Inc., 545 pp.

Niiler, P. P. (1968). On the internal tidal motions in the Florida Straits. *Deep-Sea Res.* 15, 113-123.

Orlanski, I. (1971). Energy spectrum of small-scale internal gravity waves. *J. Geophys. Res.* 76(24), 5829-5835.

Orlanski, I. (1972). On the breaking of standing internal gravity waves. *J. Fluid Mech.* 54(4), 577-598, 5 plates.

Orlanski, I. and K. Bryan (1969). Formation of the thermocline step structure by large-amplitude internal gravity waves. *J. Geophys. Res.* 74(28), 6975-6983.

Palm, E. (1958). Two-dimensional and three-dimensional mountain waves. *Geofys. Publ.* 20(3), 1-25.

Pao, Y.-H. and A. Goldburg (Eds.) (1969). *Clear air turbulence and its detection.* New York: Plenum Press, 542 pp.

Parrish, D. P. and P. P. Niiler (1971). Topographic generation of long internal waves in a channel. *Geophys. Fluid Dyn.* 2(1), 1-29.

Perkins, H. (1972). Inertial oscillations in the Mediterranean. *Deep-Sea Res.* 19(4), 289-296.

Perkins, H. (1973). Observed effects of mean shear on inertial oscillations. *EOS Trans. Amer. Geophys. Un.* 54(4), 307 (abstract).

Perry, R. B. and G. R. Schimke (1965). Large-amplitude internal waves observed off the northwest coast of Sumatra. *J. Geophys. Res.* 70(10), 2319-2324.

Peters, A. S. and J. J. Stoker (1960). Solitary waves in liquids having non-constant density. *Comm. Pure Appl. Math.* 13(1), 115-164.

Petrie, B. (1973). Internal tides on the continental slope. *EOS Trans. Amer. Geophys. Un.* 54(4), 316 (abstract).

Pettersson, H. (1916). Bewegungen des Tiefenwassers an der Küste von Bohuslän im November 1915. *Ann. Hydrogr. u. Marit. Meteorol.* 44, 270-274.

Phillips, O. M. (1960). On the dynamics of unsteady gravity waves of finite amplitude. Part 1. *J. Fluid Mech.* 9, 193-217.

Phillips, O. M. (1963). Energy transfer in rotating fluids by reflection of inertial waves. *Phys. Fluids* 6(4), 513-520.

Phillips, O. M. (1966). *Dynamics of the upper ocean.* London: Cambridge Univ. Press, 261 pp.

Phillips, O. M. (1967). The generation of clear air turbulence by the degradation of internal waves, pp. 130-138. In, Proc. Intern. Coll. Atmos. Turbulence and Radio Wave Propagation, Moscow, 1965.

Phillips, O. M. (1968). The interaction trapping of internal gravity waves. *J. Fluid Mech.* 34, 407-416.

Phillips, O. M. (1971). On spectra measured in an undulating layered medium. *J. Phys. Oceanogr.* 1, 1-6.

Phillips, O. M. (1972). Turbulence in a strongly stratified fluid --is it unstable? *Deep-Sea Res.* 19, 79-81.

Phillips, O. M., W. K. George, and R. P. Mied (1968). A note on the interaction between internal gravity waves and currents. *Deep-Sea Res.* 15, 267-273.

Piip, A. T. (1969). Internal waves in and below the Sargasso Sea thermocline. *EOS Trans. Amer. Geophys. Un.* 50(4), 188 (abstract).

Pochapsky, T. E. (1963). Measurement of small-scale oceanic motions with neutrally-buoyant floats. *Tellus* 25, 352-362.

Pochapsky, T. E. (1966). Measurements of deep water movements with instrumented neutrally buoyant floats. *J. Geophys. Res.* 71(10), 2491-2504.

Pochapsky, T. E. and F. D. Malone (1972). Spectra of deep vertical temperature profiles. *J. Phys. Oceanogr.* 2(4), 470-475.

Pollard, R. T. (1970). On the generation by winds of inertial waves in the ocean. *Deep-Sea Res.* 17, 795-812.

Pollard, R. T. (1972). Properties of near-surface inertial oscillations. Woods Hole Oceanogr. Inst., WHOI Contrib. 2736, 44 pp.

Pollard, R. T. and R. C. Millard (1970). Comparison between observed and simulated wind-generated inertial oscillations. *Deep-Sea Res.* 17, 813-821.

Polyanskaya, V. A. (1969). The generation of subsurface waves by a plane pressure front moving at constant velocity. *Izv. Atmos. Ocean. Phys.* 5(10), 608-611.

Proni, J. R. and J. R. Apel (1973). Observations of internal waves using a directional high frequency transducer. *EOS Trans. Amer. Geophys. Un.* 54(4), 323 (abstract).

Proudman, J. (1953). *Dynamical oceanography.* New York: J. Wiley and Sons, Inc., 409 pp.

Radok, J. R., W. H. Munk, and J. Isaacs (1967). A note on mid-ocean internal tides. *Deep-Sea Res.* 14, 121-124.

Rarity, B. S. H. (1969). A theory of the propagation of internal gravity waves of finite amplitude. *J. Fluid Mech.* 39(3), 497-509.

Rattray, M., Jr. (1957). Propagation and dissipation of long internal waves. *Trans. Amer. Geophys. Un.* 38, 495-500.

Rattray, M., Jr. (1960). On the coastal generation of internal tides. *Tellus* 12, 54-62.

Rattray, M., Jr., J. G. Dworski, and P. Kovala (1969). Generation of long internal waves at the continental slope. *Deep-Sea Res.* 16(suppl.), 179-196.

Regal, R. R. and C. I. Wunsch (1973). M_2 tidal currents in the western north Atlantic. *Deep-Sea Res.* 20(5), 493-502.

Reid, J. L., Jr. (1956). Observations of internal tides in October 1950. *Trans. Amer. Geophys. Un.* 37(3), 278-286.

Reid, J. L., Jr. (1958). A comparison of drogue and GEK measurements in deep water. *Limnol. Oceanogr.* 3, 160-165.

Reid, J. L., Jr. (1962). Observations of inertial rotation and internal waves. *Deep-Sea Res.* 9, 283-289.

Revelle, R. (1939). Sediments of the Gulf of California. *Bull. Geolog. Soc. Amer.* 50, 1929 (abstract).

Roberts, J. (1973). "Internal Gravity Waves in the Ocean: Bibliography and Keyword Index." Inst. Marine Sci., Univ. Alaska, Rept. R73-4, 425 pp.

Robinson, R. M. (1969). The effects of a vertical barrier on internal waves. *Deep-Sea Res.* 16, 421-429.

Robinson, R. M. (1970). The effects of a corner on a propagating internal gravity wave. *J. Fluid Mech.* 42(2), 257-267.

Rooth, C. G. (1971). A weakly diffusive thermocline model. *EOS Trans. Amer. Geophys. Un.* 52(4), 233 (abstract).

Rooth, C. G. and W. Düing (1971). On the detection of "inertial" waves with pycnocline followers. *J. Phys. Oceanogr.* 1, 12-16.

Rudnick, P. and J. D. Cochrane (1951). Diurnal fluctuations in bathythermograms. *J. Mar Res.* 10(3), 257-262.

Sabinin, K. D. (1966). Connection of short-period internal waves with the vertical density gradient in the sea. *Izv. Atmos. Ocean. Phys.* 2(8), 527-532.

Sabinin, K. D. (1969). Determination of the parameters of internal waves using data from a towed chain of thermistors. *Izv. Atmos. Ocean. Phys.* 5(2), 113-115.

Sabinin, K. D. and V. A. Shulepov (1965). Short period internal waves of the Norwegian Sea. *Oceanology* 5, 60-69.

Sabinin, K. D. and V. A. Shulepov (1970). Some results of studies on internal tidal waves in the tropic zone of the Atlantic (in Russian, with English abstract). *Izv. Akad. Nauk SSSR. Fiz. Atmos. Okeana* 6(2), 189-197.

Saint-Guily, B. (1970). On internal waves. Effects of the horizontal component of the earth's rotation and of a uniform current. *Dtsch. Hydrogr. Zeit.* 23(1), 16-23.

Sandstrom, H. (1969). Effect of topography on propagation of waves in stratified fluids. *Deep-Sea Res.* 16(5), 405-410.

Sandström, J. W. (1908). Dynamische Versuche mit Meerwasser. *Ann. Hydrogr. u. Marit. Meteorol.* 36, 6-23.

Schooley, A. H. and B. A. Hughes (1972). An experimental and theoretical study of internal waves generated by the collapse of a two-dimensional mixed region in a density gradient. *J. Fluid Mech.* 51(1), 159-175.

Schooley, A. H. and R. W. Stewart (1963). Experiments with a self-propelled body submerged in a fluid with a vertical density gradient. *J. Fluid Mech.* 15, 83-96.

Schott, F. (1970). Long thermocline waves in the North Sea. *Univ. Liège Cah. Méc. Math.* 26, 89-91.

Schott, F. (1971a). On horizontal coherence and internal wave propagation in the North Sea. *Deep-Sea Res.* 18, 291-307.

Schott, F. (1971b). Spatial structure of inertial-period motions in a two-layered sea, based on observations. *J. Mar. Res.* 29(2), 85-102.

Schott, F. and J. Willebrand (1973). On the determination of internal wave directional spectra from moored instruments. *J. Mar. Res.* 31, 116-134.

Schubert, O. V. (1939). Bericht über die zweite Teilfahrt der Deutschen Nordatlantischen Expedition „Meteor" Januar-Juli 1938. *Ann. Hydro. u. Marit. Meteorol.* 67 (Jan.-Beiheft.), 11-18.

Schule, J. J., Jr. (1952). "Effects of Weather upon the Thermal Structure of the Ocean." U.S. Hydrographic Office, Washington, D.C., Prog. Rept. 1. H. O. Miscellaneous 15360, 81 pp.

Scorer, R. S. (1950). The dispersion of a pressure pulse in the atmosphere. *Proc. Roy. Soc. London, Ser. A* 201, 137-157.

Scotti, R. S. and G. M. Corcos (1972). An experiment on the stability of small disturbances in a stratified free shear layer. *J. Fluid Mech.* 52(3), 499-528.

Seiwell, H. R. (1937). Short period vertical oscillations in the western basin of the North Atlantic. *Pap. Phys. Oceanogr. Meteorol. Woods Hole Oceanogr. Inst.* 5(2), 1-44.

Seiwell, H. R. (1939). Daily temperature variations in the western North Atlantic. *J. Cons. Perm. Intern. Explor. Mer* 14(3), 357-369.

Seiwell, H. R. (1942). An analysis of vertical oscillations in the southern North Atlantic. *Proc. Amer. Phil. Soc.* 85(2), 136-158.

Selitskaia, E. S. (1957). On the problem of internal waves (in Russian). *Meterol. i Gidrol. (Leningrad)* 6, 35-39.

Shand, J. A. (1953). Internal waves in Georgia Strait. *Trans. Amer. Geophys. Un.* 34, 849-856.

Shen, M. C. (1969). Stationary nonlinear waves in a stratified fluid with respect to a rotating cylindrical system. *Phys. Fluids* 12, 1961-1967.

Shlyamin, B. A. (1957). On internal waves in the southern part of the Atlantic Ocean (in Russian). *Izvestiya Geograficheskoe Obshchestvo SSSR* 85, 470-474.

Shonting, D. H., F. de Strobel, A. de Haen, L. Toma, and R. Maggiora (1972). Thermistor buoy observations in the Mediterranean. *EOS trans. Amer. Geophys. Un.* 53(4), 426 (abstract).

Siedler, G. (1968). Schichtungs- und Bewegungsverhältnisse am Südausgang des Roten Meeres. *„Meteor" Forsch.-Ergebnisse A, Meteorol. Hydrol.* 4.

Siedler, G. (1971). Vertical coherence of short-periodic current variation. *Deep-Sea Res.* 18, 179-191.

Simmons, W. E. (1969). Variational method for weak resonant wave interactions. *Proc. Roy. Soc. London. Ser. A* 309, 551-575.

Smith, R. W. (1972). An instability of internal gravity waves. *J. Fluid Mech.* 52(2), 393-399.

Southard, J. B. and D. A. Cacchione (1972). Experiments on bottom sediment movement by breaking internal waves. In, *Shelf sediment transport*; Swift, Duane and Pilkey (Eds.), Stroudsburg, PA: Dowden, Hutchinson & Ross, Inc., Chap. 4, pp. 83-97.

Spiegel, E. A. and G. Veronis (1960). On the Boussinesq approximation for a compressible fluid. *J. Astrophys.* 131, 442-447.

Sretinskii, L. N. (1959). On the wave resistance of ships in the presence of internal waves (in Russian). *Izv. Akad. Nauk SSSR, Otdelenie Teknicheskikh* 1, 56-63.

Stern, M. E. and J. S. Turner (1969). Salt fingers and convecting layers. *Deep-Sea Res.* 16, 497-511.

Stevenson, T. N. (1968). Some two-dimensional internal waves in a stratified fluid. *J. Fluid Mech.* 33(4), 715-720.

Stevenson, T. N. (1969). Axisymmetric internal waves generated by a traveling oscillating body. *J. Fluid Mech.* 35(2), 219-224.

Stevenson, T. N. and N. H. Thomas (1969). Two-dimensional internal waves generated by a traveling oscillating cylinder. *J. Fluid Mech.* 36(3), 505-511.

Summers, H. J. and K. O. Emery (1963). Internal waves of tidal period off Southern California. *J. Geophys. Res.* 68(3), 827-839.

Takano, K. (1969). Houle interne de second ordre (in French, with English and Japanese abstracts). *La Mer (Bulletin de la Société Franco-Japanaise d'Océanographie)* 7(2), 119-125.

Takano, K. and N. Iida (1969). Generation of internal waves by an abrupt change of depth (in English, with French and Japanese abstracts). *La Mer (Bulletin de la Société Franco-Japonaise d'Océanographie)* 7(2), 150-160.

Tareev, B. A. (1963). Internal waves in an ocean of nonuniform density. *Doklady Akad. Nauk SSSR* 149(4), 827-830. (English translation by Amer. Geol. Inst., *Geophys.* 5-7.)

Tareev, B. A. (1965). Internal baroclinic waves in a flow around irregularities on the ocean floor and their effect on sedimentation processes. *Oceanology* 5(1), 31-37.

Taylor, G. I. (1931). Internal waves and turbulence in a fluid of variable density. *Rapports et Procès-Verbaux, Intern. Council Expl. Sea (Copenhagen)* 76, 35-43.

Testud, J. (1970). Gravity waves generated during magentic sub-storms. *J. Atmos. Terr. Phys.* 32, 1793-1805.

Thomas, N. H. and T. N. Stevenson (1972). A similarity solution for viscous internal waves. *J. Fluid Mech.* 54(3), 495-506.

Thorade, H. (1928). Gezeitenuntersuchungen in der deutschen Bucht der Nordsee. *Arch. Dtsch. Seewarte, Hamburg* 46(5), 1-84.

Thorpe, S. A. (1966). On wave interactions in a stratified fluid. *J. Fluid Mech.* 24(4), 737-751.

Thorpe, S. A. (1968a). A method of producing a shear flow in a stratified fluid. *J. Fluid Mech.* 32(4), 693-704.

Thorpe, S. A. (1968b). On standing internal gravity waves of finite amplitude. *J. Fluid Mech.* 32(3), 489-528.

Thorpe, S. A. (1968c). On the shape of progressive internal waves. *Phil. Trans. Roy. Soc. London, Ser. A* 263(1145), 563-614.

Thorpe, S. A. (1969a). Experiments on the instability of stratified shear flows--immiscible fluids. *J. Fluid Mech.* 39(1), 25-48.

Thorpe, S. A. (1969b). Experiments on the stability of stratified shear flows. *Radio Sci.* 4(12), 1327-1332.

Thorpe, S. A. (1970). "Internal Gravity Waves--A Report on the Advances in the Understanding of Internal Gravity Waves Since 1964, and Their Possible Relation to Clear Air Turbulence." Prepared for the Atmospheric Environment Committee Meeting on 27 January 1970, 28 pp.

Thorpe, S. A. (1971a). Asymmetry of the internal seiche in Loch Ness. *Nature* 231(5301), 306-308.

Thorpe, S. A. (1971b). Experiments on the instability of stratified shear flows--miscible fluids. *J. Fluid Mech.* 46(2), 299-319.

Thorpe, S. A. (1973). Turbulence in stably stratified fluids: a review of laboratory experiments. *Boundary-Layer Meteorol.* 5, 95-119.

Thorpe, S. A., A. Hall, and I. Crofts (1972). The internal surge in Loch Ness. *Nature* 237(5350), 96-98.

Tolstoy, I. (1963). The theory of waves in stratified fluids including the effects of gravity and rotation. *Rev. Mod. Phys.* 35(1), 207-230.

Tolstoy, I. and J. Lau (1971). Generation of long internal gravity waves in waveguides by rising buoyant air masses and other sources. *Geophys. J. Roy. Astron. Soc.* 26(1-4), 295-310.

Tomczak, M., Jr. (1966). Winderzeugte interne Wellen, insbesondere im Periodenbereich oberhalb der Trägheitsperiode. *Dtsch. Hydrogr. Zeit.* 19, 1-21.

Tomczak, M. Jr. (1967a). Über den Einfluss fluktuierender Windfelder auf ein stetig geschichtetes Meer. *Dtsch. Hydrogr. Zeit.* 20(3), 101-129.

Tomczak, M., Jr. (1967b). Über Eigenwerte freier und Resonanz erzwungener interner Wellen in einem stetig geschichteten flachen Meer. *Dtsch. Hydrogr. Zeit.* 20(5), 218-232.

Tomczak, M., Jr. (1968). Über interne Wellen in der Nähe der Trägheitsperiode. *Dtsch. Hydrogr. Zeit.* 21(4), 145-151.

Tomczak, M., Jr. (1969). Über interne Trägheitsbewegungen in der Westlichen Ostsee. *Dtsch. Hydrogr. Zeit.* 22(4), 158-162.

Tominaga, M. (1970). Finite vertical oscillation of fluid parcel in a heterogeneous, incompressible sea water. Part II. Observational results in the sea. *J. Oceanogr. Soc. Japan.* 26(5), 267-270.

Townsend, A. A. (1964). Natural convection in water over an ice surface. *Quart. J. Roy. Met. Soc.* 90, 248-259.

Townsend, A. A. (1965). Excitation of internal waves by a turbulent boundary layer. *J. Fluid Mech.* 22(2), 241-252.

Townsend, A. A. (1966). Internal waves produced by a convective layer. *J. Fluid Mech.* 24, 307-319.

Ufford, C. W. (1945). "Internal Waves off San Diego, California." Univ. Calif. Division of War Res., USNEL, San Diego, Calif., Rept. M 290 (unclassified), 8 pp., 21 figs.

Ufford, C. W. (1946). "Internal Waves Measured with Three Thermocouples." Univ. Calif. Division of War Res., USNEL, San Diego, Calif., Rept. M 406 (unclassified), 13 pp., 20 figs.

Ufford, C. W. (1947). Internal waves measured at three stations. *Tran. Amer. Geophys. Un.* 28, 87-95.

University of California Division of War Research (1942). "A Laboratory Study of Surface and Internal Wave Motion." U.S. Navy Radio and Sound Laboratory, San Diego, Calif., UCDWR Rept. U3 (unclassified), 18 pp., 8 figs.

Valdez, V. (1960). Internal waves on a echo sounder record. *Deep-Sea Res.* 7, 148.

Van Leer, J. C. (1971). Ph.D. thesis. "Shear of Small Vertical Scale Observed in the Permanent Oceanic Thermocline." MIT, and Woods Hole Oceanogr. Inst., 209 pp.

Van Leer, J. C., W. Düing, R. Erath, E. Kennelly, and A. Speidel (1974). The cyclesonde: an unattended vertical profiler for scalar and vector quantities in the upper ocean. *Deep-Sea Res.* 21, 385-400.

Verber, J. L. (1964). The detection of rotary currents and internal waves in Lake Michigan, pp. 382-389. In, Proc. 7th Conf. Great Lakes Res., Univ. Michigan, Great Lakes Res. Div., Publ. 11.

Vergeiner, I. (1971). An operational linear lee wave model for arbitrary basic flow and two-dimensional topography. *Quart. J. Roy. Met. Soc.* 97(411), 30-60.

Veronis, G. (1967). Analogous behaviour of rotating and stratified fluids. *Tellus* 19, 620-634.

Voit, S. S. and B. I. Sebekin (1969). The diffraction of nonsteady surface and subsurface waves. *Izv. Atmos. Ocean. Phys.* 5(2), 94-97.

Voorhis, A. D. (1968). Measurements of vertical motion and the partition of energy in the New England slope water. *Deep-Sea Res.* 15, 599-608.

Voorhis, A. D. and H. T. Perkins (1966). The spatial spectrum of short-wave temperature fluctuations in the near-surface thermocline. *Deep-Sea Res.* 13, 641-654.

Walker, L. R. (1973). Interfacial solitary waves in a two-fluid medium. *Phys. Fluids* 16(11), 1796-1804.

Wang, Y.-C. (1969). The interaction of internal waves with an unsteady non-uniform current. *J. Fluid Mech.* 37(4), 761-771.

Warner, J. L. (1972). On the effect of topography on the propagation of internal waves. *Rapports et Procès-Verbaux, Intern. Council Expl. Sea (Copenhagen)* 162, 80-84.

Watson, E. R. (1904). Movements of the waters of Loch Ness, as indicated by temperature observations. *Geogr. J. London* 24, 430-437.

Webster, F. (1963). A preliminary analysis of some Richardson current meter records. *Deep-Sea Res.* 10, 389-396.

Webster, F. (1968). Observations of inertial-period motions in the deep sea. *Rev. Geophys.* 6(4), 473-490.

Webster, F. (1970). Lectures. *Univ. Liège Cah. Méc. Math.* 26, 20-53.

Webster, F. and N. P. Fofonoff (1967). "A Compilation of Moored Current Meter Observations, Vol. 3." Woods Hole Oceanogr. Inst., WHOI Rept. 67-66, 105 pp. (Cited by Webster, 1968.)

Wedderburn, E. M. (1909a). Dr. O. Pettersson's observations on deep water oscillations. *Proc. Roy. Soc. Edinburgh* 29, 602-606.

Wedderburn, E. M. (1909b). Temperature observations in Loch Garry. *Proc. Roy. Soc. Edinburgh* 29, 98-135.

Wedderburn, E. M. (1910). Current observations in Loch Garry. *Proc. Roy. Soc. Edinburgh* 30, 312-323.

Wedderburn, E. M. (1912). Temperature observations in Loch Earn, with a further contribution to the hydrodynamical theory of the temperature seiches. *Phil. Trans. Roy. Soc. Edinburgh* 48, 629-695.

Wedderburn, E. M. and W. Watson (1909). Observations with a current meter in Loch Ness. *Proc. Roy. Soc. Edinburgh* 29, 619-647.

Wedderburn, E. M. and A. M. Williams (1911). The temperature seiche. *Trans. Roy. Soc. Edinburgh* 47, 619-642.

Weigand, J. G., H. G. Farmer, S. J. Prinsenberg, and M. Rattray, Jr. (1969). Effects of friction and surface tide angle of incidence on the coastal generation of internal waves. *J. Mar. Res.* 27(2), 241-259.

Weissman, M. A. (1972). Nonlinear development of the Kelvin-Helmholtz instability. *EOS Trans. Amer. Geophys. Un.* 53(4), 417 (abstract).

Weston, D. E. and W. W. Reay (1969). Tidal period internal waves in a tidal stream. *Deep-Sea Res.* 16(5), 473-478.

White, R. A. (1967). Vertical structure of temperature fluctuations within an oceanic thermocline. *Deep-Sea Res.* 14, 613-623.

Whitham, G. B. (1961). Group velocity and energy propagation for three-dimensional waves. *Comm. Pure Appl. Math.* 14, 675-691.

Woods, J. D. (1968). Wave induced shear instability in the summer thermocline. *J. Fluid Mech.* 32(4), 791-800.

Woods, J. D. (1969a). CAT under water. *Weather* 23, 224-235.

Woods, J. D. (1969b). On Richardson's number as a criterion for laminar-turbulent-laminar transition in the ocean and atmosphere. *Radio Sci.* 4, 1289-1298.

Woods, J. D. and G. G. Fosberry (1967). Structure of the thermocline, pp. 5-18. In, Underwater Association Rept. 1966-1967, Underwater Association, London.

Wu, J. (1969). Mixed region collapse with internal wave generation in a density-stratified medium. *J. Fluid Mech.* 35(3), 531-544.

Wunsch, C. I. (1968). On the propagation of internal waves up a slope. *Deep-Sea Res.* 25, 251-258.

Wunsch, C. I. (1969). Progressive internal waves on slopes. *J. Fluid Mech.* 35, 131-144.

Wunsch, C. I. and J. Dahlen (1970a). Internal waves and mixing processes at Bermuda. *EOS Trans. Amer. Geophys. Un.* 51(4), 314 (abstract).

Wunsch, C. I. and J. Dahlen (1970b). Preliminary results of internal wave measurements in the main thermocline at Bermuda. *J. Geophys. Res.* 75(30), 5899-5908.

Wunsch, C. I. and R. Hendry (1972). Array measurements of the bottom boundary layer and the internal wave field on the continental slope. *Geophys. Fluid Dyn.* 4, 101-145.

Yampolskii, A. D. (1960). On inertial motions in the Black Sea, from observations at a multiday anchor station (in Russian). *Trudy Akad. Nauk SSSR, Institut Okeanologii* 39, 111-126. (Cited by Krauss, 1966b.)

Yampolskii, A. D. (1962). On internal waves in the northeast Atlantic (in Russian). *Trudy Akad. Nauk SSSR, Institut Okeanologii.* 56, 229-240.

Yearsley, J. R. (1966). "Internal Waves in the Arctic Ocean." Lamont Geol. Observ., Columbia Univ., Palisades, New York., Tech. Rept. 5, 63 pp.

Yih, C.-S. (1960). Gravity waves in a stratified fluid. *J. Fluid Mech.* 8, 481-508.

Yih, C.-S. (1969). Stratified flows. In, *Ann. Rev. Fluid Mech.*, Vol. 1, W. R. Sears (Ed.), Palo Alto, Calif.: Annual Reviews, pp. 73-110.

Young, J. A. and C. W. Hirt. (1972). Numerical calculation of internal wave motions. *J. Fluid Mech.* 56(2), 265-276.

Zalkan, R. L. (1969). Internal waves of tidal periodicity. *J. Geophys. Res.* 74(13), 3434-3435.

Zalkan, R. L. (1970). High frequency internal waves in the Pacific Ocean. *Deep-Sea Res.* 17, 91-108.

Zeilon, N. (1912). On tidal boundary-waves and related hydrodynamical problems. *Kungl. Svenska Vetensk.-Akad. Handl.* 47(4), 1-46.

Zeilon, N. (1934). Experiments on boundary tides. *Medd. Goteborgs Hogskolas Oceanogr. Inst. B* 3(10), 1-8.

Ziegenbein, J. (1969). Short internal waves in the Strait of Gibraltar. *Deep-Sea Res.* 16(5), 479-487.

Ziegenbein, J. (1970). Spatial observations of short internal waves in the Strait of Gibraltar. *Deep-Sea Res.* 17, 867-875.

Zubov, N. N. (1932). Hydrographical investigations in the southwestern part of the Barents Sea during the summer of 1928 (in Russian). *Trudy Moscow Gosudarstennii Okeanograficheskii Institut* 2(4), 3-79.

AUTHOR INDEX

Numbers that are underlined refer to the page on which the complete reference is cited.

Y

Z

SUBJECT INDEX

I

J, K

L